Proteas

THE BIRTH OF A WORLDWIDE INDUSTRY

A historical overview of the commercialisation
of Proteas and Cape Greens

Maryke Middelmann

Copyright © 2012 by Maryke Middelmann.

Library of Congress Control Number: 2011962076
ISBN: Hardcover   978-1-4691-3319-5
                 Softcover   978-1-4691-3318-8
                 Ebook   978-1-4691-3320-1

All rights reserved. No part of this book may be reproduced or transmitted in any form or by any means, electronic or mechanical, including photocopying, recording, or by any information storage and retrieval system, without permission in writing from the copyright owner.

This book was printed in the United States of America.

To order additional copies of this book, contact:
Xlibris Corporation
0-800-644-6988
www.xlibrispublishing.co.uk
Orders@xlibrispublishing.co.uk

Contents

Preface..6
Early History...7
Early Commercial Publications...13
Conservation and Organisations..18
South African Wild Flower Growers Association (SAWGRA).............................23

Industry Developments 1970 - 1979
     South Africa..47
     International Developments..66
     The International Protea Association (IPA)....................................74

Industry Developments 1980 - 1985
     South Africa..77
     International Developments..92
     The International Protea Working Group (IPWG).................................95

Industry Developments 1986 - 1989
     South Africa...109
     International Developments...126

Industry Developments 1990 - 1994
     South Africa...137
     International Developments...148

Industry Developments 1995 - 1996
     South Africa...158
     International Developments...165

Industry Developments 1997 - 1999
     South Africa...168
     International Developments...175

Industry Developments 2000 - 2005
     South Africa...179
     International Developments...196

2005 Onwards
    The Break-up of SAPPEX 202
    International Developments...208
    Other International News 209

Appendix A: South African Proteas..211

Appendix B: Some Aspects Of The Problem Of Preserving
    The Indigenous Flora Of S.A. 213

Appendix C: The Story Of SAPPEX 225

Appendix D: The Birth Of The Dried Flower Industry
    In South Africa 229

Appendix E: South African Protea Producers
    And Exporters Association...232

Appendix F: Survey Wildflower Production, 1981 237

Appendix G: Marie Vogts Appointed As Researcher At Fruit
    And Food Technology Institute (FFTRI) 244

Appendix G: Research Into The South African Proteaceae 247

Appendix I: Development And Structure
    Of The South African Protea Industry 251

Appendix J: The Conservation Of Genetic Resources
    In The Southern African Proteaceae 265

Appendix K: Safari Sunset 274

Dedication

This book is dedicated to the memory of
Walter and Ruth Middelmann
who started it all for us.

Preface

Walter Middelmann was a collector of note. After his death, we had a mammoth task to sort some semblance of order into all the papers, documents, memorabilia and diaries he had kept from an early age. Together with papers from his life in Germany and his new life in South Africa, there were literally boxes and boxes full of Protea Industry papers, as well as papers on other organisations, which held his interest. Having all this documentation, and having spent a large chunk of my personal life on Industry affairs, I felt inspired to write it all down for the benefit of industry members and others who might be interested.

I particularly want to thank Barrie Gibson for his encouragement and for proofreading the first draft. Of course, I must thank Robert, my husband, who encouraged me from day one and who was his ever-patient self when I got carried away!

Maryke Middelmann
Botrivier, South Africa
2011

Early History

It is commonly known that Proteaceae occur naturally in South Africa and Australia. Less known are the Proteaceae of South America: *Embothrium coccineum, Gevuina avellana, Lomatia ferruginea* and it certainly would be interesting to find some early literature on these species.

The Middelmann family's history of involvement in proteas started when Walter and Ruth Middelmann, both immigrants from Germany before the Second World War, were married. As a young bride, Ruth became interested in the proteas growing above their home on the slopes above Clifton on the back of Lion's Head in Cape Town. When, in 1948, they bought a piece of 'useless' mountain side 100 km outside Cape Town as a weekend retreat, Ruth turned her early attempts at protea propagation into reality and that is more or less where this story starts.

A few years down the line in 1966, with quite a bit of experience gained about the Cape Flora, Walter Middelmann gave a talk in Australia on developments at the Cape. In it he referred to different phases in the history of wild flowers at the Cape, namely the:

- *Historical/Botanical Phase* (1605-1910);
- *Intellectual Phase* (1910-1920): Local botanists formed organisations such as the Wild Flower Protection Society and the Botanical Society to prevent destruction not only by street sellers, but by 'progress';
- *Restrictive phase* (1920-1945): To stop the indiscriminate plucking whether for gain or for fun, protection of flowers leading to Education and Publicity through shows, lectures, and creating interest in the use of indigenous flora;
- *Small-scale commercial phase* (1940-1959): Attempts to define 'cultivation' and popularity of wild plants spreading throughout the country as a result of the efforts of pioneers like Frank Batchelor, the Middelmanns and others. At this time the first of Marie Vogts' books was published;
- *The Commercial Phase* (1960 onwards): Making Proteas pay by planting orchard-like in rows, easing of legislation to make

'utilisation' possible and the ideology of preservation of natural sites being in the interest of the owner.

Not much detail is available to me on these different phases, particularly the Intellectual and Restrictive Phases and that could be an interesting period to research for anyone so inclined. This book concentrates on the Commercial Phase and organisations and events impacting on the commercial sector.

The history of the Protea Industry worldwide would not be complete without at least a reference to the historical/botanical phase and the first European explorers who 'discovered' these new floral treasures. For this it is necessary to refer to some old botanical books from that era. South Africa has contributed significantly to world horticulture by providing plants such as Gladiolus, Kniphofia, Pelargonium, Freesia, Tritonia, Sparaxis, Gerbera, Strelizia, Plumbago, Tecomaria, Agapanthus, Streptocarpus, Nemesia and a host of others, particularly the Proteaceae, although these were cultivated much later than the others. According to Gail Littlejohn, head of the ARC Breeding Programme, this was probably due to the fact that they were quite difficult to raise.

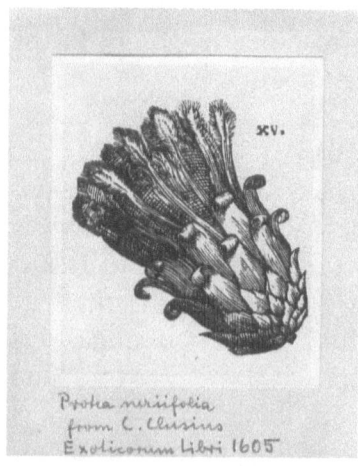

Protea neriifolia, Exoticorum Libri 1605 (Middelmann library)

The first ever reference to a protea is found in *Clusius' Exoticorum Libri Decem*, published in Antwerpen in 1605. This early book has a woodcut of a *Protea neriifolia* flower, which provided an astonished Europe with the *first depiction of a South African plant*. Clusius provided a full description that he referred to as a 'thistle'. Although he reported that it was collected in Madagascar during a trading expedition to Java, its locality was clearly incorrectly recorded and he must have collected it somewhere along the Cape coast during a stop for fresh water. The fresh flower, having been taken back to Europe, was quite dried out by the time it arrived and was sketched. Walter Middelmann, who was to establish Honingklip Dryflowers in 1964, liked to say that it probably was South Africa's first dried flower export!

By 1687, Paul *Hermann*, professor at Leiden, in his *Horti Academici* already listed thirty-four Cape plants.

Volumes of illustrated books (*Curtis Botanical Magazine*) from the 1700s referred to proteas already. This 'magazine' was later published under the title 'Kew Magazine'. The 1779 magazine featured a *protea mellifera* (now *P. repens*) or Sugarbush, which was reproduced by the S.A. Library as a Christmas card a number of years ago.

Swedish travelers joined in the hunt for plants. Both *Thunberg* and *Carl Gustav's 'Ostindiska Resa'* of 1773 (East India Travel) mention visits to the Cape. Peter *Bergius* described Cape Flora in 1767.

Francis Masson (1741-1805), a native of Aberdeen, Scotland, was brought up to be a gardener, but decided to seek his fortune in the south and became an under-gardener to the Royal gardens at Kew. In early 1772, he was appointed botanical collector for Kew at the Cape of Good Hope. He stayed at the Cape for two and a half years, and became acquainted with the Swedish botanist Carl Pehr Thunberg. Masson wrote as preface to his book on Stapelias:

> *The curious productions of the Cape had been too much neglected until the year 1771, when Captain Cook returned from his first voyage round the globe, and landed the naturalists who accompanied him at the Cape Town; they were much gratified by the treasures they met with, and in consequence of the observations they then made, Sir Joseph Banks, on his return to England, suggested to His Majesty the idea of sending a person, professionally a gardener, to the Cape, to collect seeds and plants for the Royal botanic gardens at Kew; His Majesty was graciously pleased to adopt the plan, though at that time so little approved by the public, that no one but myself chose to undertake the execution of it. I sailed for the Cape in the beginning of 1772 and remained there two years and a half.*

After some further expeditions he spent another ten years at the Cape.

No wonder therefore that the first proteas to be grown away from their native South Africa flowered at the Royal Botanic Gardens, Kew, over 200 years ago. The seed had been collected by Francis Masson and the plants were tended by his sponsor, Sir Joseph Banks and William Aiton, the King's chief gardener at Kew. By 1810, there were twenty-three species of Protea in cultivation at Kew and this was the beginning of a horticultural fashion among the wealthy that was to last until the 1830s.

Thunberg (1743-1818), a pupil of Linnaeus, worked his way to South Africa as assistant surgeon. He reached the Cape in April 1772 and stayed for about three years. His visit almost exactly coincided with the first visit of Francis Masson. His '*Flora Capensis*', based on his own collections, mentions no less than 3100 species. He is considered the father of South African Botany. On his return to Sweden he was appointed to the chair at Upsala University, following in the footsteps of Linnaeus.

W.H. Harvey's *Genera of South African Plants* was the first locally produced book on the Cape Flora in 1838.

Joseph Knight's information in regard to the cultivation of proteaceae covers only fourteen pages and it is most remarkable that in later years so little attention was paid to his recommendations. He wrote for example that at all times and at all ages fresh air is an absolute essential for these plants and that they should be kept in the airiest section of the greenhouse during the European winter, otherwise they would definitely 'damp off'. When greenhouse heating improved, the protea did not survive. Knight also stressed the low temperature at which the seed of the plant germinates. He told of a friend of his who was given some seeds of the silver tree, which had been brought aboard ship to feed the turkeys. This friend had tried to germinate some of the seeds in a hotbed without success. A few had fallen on a cool damp spot and of these a number had come up. Knights further stated that he thought the whole family had a dislike to be sown in artificial heat. Today we consider it best to sow seed in autumn and winter in a temperate climate. Joseph Knight's book, *On the order of the plants belonging to the natural order of 'Proteeae'*, which appeared in 1809 is a rare work and is an essential book to any understanding of the taxonomic history of the Proteaceae. A facsimile reprint came out in Cape Town in 1987, with an introduction by Dr. J.P. Rourke, head of the Compton Herbarium, Kirstenbosch.

Lichtenstein's 'Reisen '(travels) 1803-06, was published in Berlin. He wrote about crossing the Botrivier. He mentions a number of Everlastings in bloom in that region, including *E. sesamoides*, *E. vestitum* and *E. imbricatum*. He writes that the locals call all of them 'Sevenjaars bloemen' (seven years flowers) because after harvesting at the optimum stage, the flowers last for seven years . . . '*They are extensively used for decoration, and are taken as gifts when the farmers go to town. Europeans buy these as a fashion item to sell back in Europe. After Protea and Erica, these plants are the most numerous*'. Lichtenstein spent some time resting at Boontjieskraal between Botrivier and Caledon, a property then belonging to Field-Cornet Conrad Greeve, before continuing his journey.

Sievers-Hahn's '*Afrika*', published in 1906, also refers to Everlastings. On page 168 of this publication the author wrote: '*the flowers of the Helichrysum vestitum (now Syncarpa vestita) when their time comes, look like fresh fallen snow in the landscape. They are collected in masses and dried and then go to London and Hamburg, where they are further distributed, with many of them going to Russia to decorate the churches.*' Hence, in the trade in Germany they became well known as 'Capblumen' (Cape Flowers) — a name that is still in use today.

Much to my surprise I came across a copy of a proclamation by Lord Charles Henry Somerset regarding the conservation of 'The Boekhoo Plant', dated December 1824. The proclamation reads as follows:

Whereas I have been given to understand that the medicinal qualities of the Boekhoo (Boego — Buchu) plant are held in great estimation in Europe, and that it is probable the demand for that article may increase to an extent which may prove very beneficial to the interest of this Colony, provided the necessary measures be taken for his preservation.

And whereas it has been represented to me that the persons employed in collecting this article, not satisfied with gathering the leaves, or even cutting the shoots, of this plant, are in the habit of pulling it up entirely by the roots, or of cutting and hewing it so low down and in such a manner as to destroy the plant itself: I have deemed it necessary for the general interests of the Colony, to order and declare, and it is hereby ordered and declared accordingly, that any person who may be convicted before a competent tribunal, of tearing up the Boekhoo (Boego) plants by the roots, or of burning it, or cutting it in such a manner as to injure the further growth of the plant, shall be deemed guilty of a misdemeanour, and be fined in a penalty of not less than twenty, nor more than fifty rix-dollars for every such offence, one-third of which shall go to the informer (provided always, that the property so injured or destroyed be not the private property of individuals, and cut or pulled up or burnt by their orders): And in the event of the inability of the offender to pay the fine awarded, that he be liable to imprisonment at hard labour for a certain period not exceeding two months for every such offence.

And that no person may plead ignorance hereof, this shall be published and affixed in the usual manner.

If not the first, it certainly is a very early Conservation statement!

As late as 1911, **Dr Marloth** wrote of a single protea he saw growing in England: 'Protea cynaroides was seen by us in flower at Kew, but the shoots were thin and slender, over 10 feet high and tied up against a trellis, not stout and robust as our wild plant'.

Early Commercial Publications

There were a number of early overseas commercial publications; notably that of Edwin Smith's Clifton Nursery, Walkerville near Adelaide in 1885 in which he offered *Protea millifera* (Cape Honeysuckle) now called *Protea repens*.

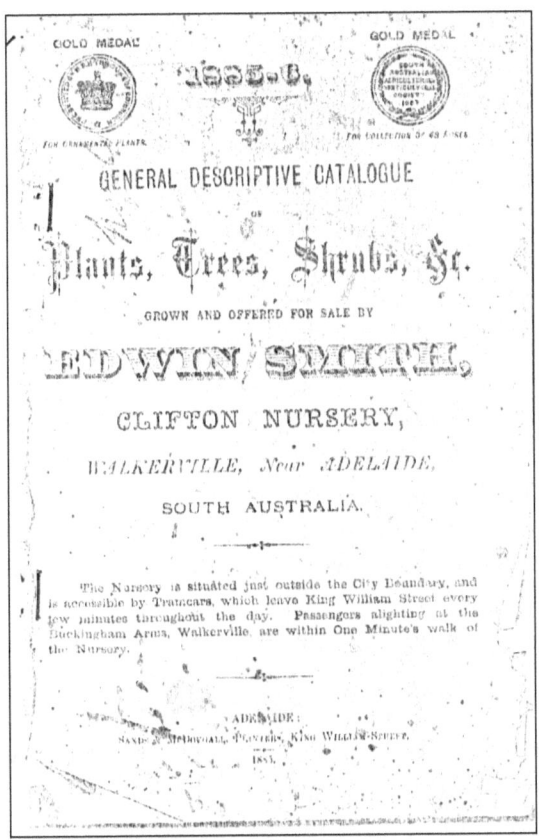

A few years later in a catalogue dated 1894/5 CF Newman & Sons lists numerous Ericas for sale, as well as *Protea mellifera* available in pink and cream under Hardy Ornamental Trees and Shrubs.

Name	Approximate ultimate height in feet.	Price per plant.
		s. d. s.
E. PROSTRANTHERA rotundifolius	4	 2
" spinulosus	3	 1
E. PROTEA mellifera – *Cape Honeysuckle* Very handsome shrub with tulip-shaped flowers of a pinkish colour.	6	... 1/6 to 2
D. PRUNUS sinensis fl. pl. alba, white	4	 1
D. " " " rosea	4	 2
E. PSIDIUM aromaticum — *yellow guava*	8	 2
" Cattleyanum — *purple guava*	8	 2
E. " pomiferum	8	 2
E. " pyriforme	8	 1
D. PUNICA granatum — *Fruiting pomegranate*	10	 1
" " fl. pl., double scarlet	10	 2
D. PYRUS. *See Cydonia.*		
D. QUERCUS coccinea	40	 3
" Ilex — *Evergreen oak*	40	... 1/6 to 2
	60	... 1/- to 1

Protea for sale. Early Australian Catalogue.
(Clifton Nursery, 1885).

In 1891, Garden and Forest, (presumably from the UK) vol 4, pages 412, 413 has an article on *Protea nana,* stating that it was grown at Kew from seed obtained from Prof MacOwen of Cape Town in 1882. At the time there were only 60 known species of Protea. With the exception of two found in Abyssinia, they were all native to South Africa.

In South Africa, awareness of the uniqueness of our flora and natural environment was already quite strong in the 1920s. Edith Stephens, a botany lecturer at UCT, was also President of the Natural History Club. One of her lectures at UCT in 1922 was on 'Life in our Vleis' (wetland). She had with her numerous samples of the wildlife, animal and plants making up this vibrant ecosystem. It was obviously a passion of hers, as she had a vast knowledge of each and every wetland on the Cape Flats. She closed her lecture with the express hope that people from different associations should lobby for protection of the vleis. An important wetland reserve on the Cape Flats now bears her name.

Commercial interest was also already receiving attention in South Africa. A number of nurseries and farmers had made the first steps towards popularizing what we now call Fynbos.

Bloem Erf Nurseries in Stellenbosch was started by Miss K.C. Stanford in 1928. The locals did not yet appreciate paying for what they could get free from nature, and her main income was from propagating and selling indigenous bulbs overseas. The Catalogue of 1946-1947 listed numerous native bulbs, seeds and plants, including the proteaceae family, for sale throughout the world. An earlier, undated, catalogue does not list proteas yet.

Protea Heights: One of the best-known growers of Proteas was Frank Batchelor. Already in 1949 he published a list of Protea plants for sale, among which were 1 Serruria, 3 Leucospermums and 7 Proteas. Together with this list he published some hints for growing these species. At that time it was commonly thought that proteas don't transplant, which of course led to the idea that it was not possible to cultivate them. The Batchelor's opened their farm to the public on each day except Mondays, Thursdays and Fridays. Admission was 2/ — (two shillings) (equal to 20c at the time when South Africa converted to Rand and cents.) Protea Heights supplied Stuttafords with plants (pricelist 1954).

(Refer to Appendix A: Gert Brits — Frank Batchelor starts protea cultivation in South Africa)

Daniel Carse of Stanford was one of the early dried flower operators who, if a newspaper cutting of 1971 is to be believed, made a fortune exporting everlasting flowers to Germany in the early 1900s. (Apparently Daniel also used to make *witblits*, a potent S.A. brandy in his own distillery.) Some interesting early publications make reference to the value of Everlastings, like this one from *Burton*:

Cape Colony for the Settler, 1903, page 27 Chapter 'A decade of Exports':

Everlasting Flowers made their mark by rising, at the starting period, from over £4000 to £12,000 in 1892 and £20,000 in 1893; fell back to about £6,500 in 1894; rose to £11,000 in the following year and again to £20,000 in 1896; receded to £11,500 in 1897, and again attained

£23,366 in 1901. *The fluctuations in this case were undoubtedly caused by the varying demands of foreign markets and the occasional limited growth of a bad season.*

And there was also the following publication by SIM: *Forest and Forest Flora of the Colony of the Cape of Good Hope, 1907:*

On the summits of the Zwartberg range the showy white everlastings flower, Helichrysum vestitum, grows abundantly. As much as 200,000 lbs weight of dry flower-heads has been sent from Caledon district in one season. In 1895 the price was 9d and that year before 1/6 per lb, but it has dropped as low as 4½ d. The flower heads are collected by children and all sorts of frail and needy people from September till the end of December, and about 2,000 flowers go to the pound weight of the product ready for market. They are used largely for church decorations and immortelle wreaths. In Russia they are much patronised by the Greek Church. The great European markets are London and Hamburg, from when they are distributed to all parts of the Continent.

Ruth and Walter Middelmann would leave their mark on the South African Fynbos Industry. As early as 1948, Ruth was already corresponding with and supplying bulbs to Kirstenbosch. There is a letter from N.S. Pillans of Bolus Herbarium, in which he thanks Ruth for bulbs she donated for the Kirstenbosch Garden, which after flowering would be dried for the Herbarium. Ruth had started her protea garden on the slopes above Clifton where she and Walter had settled after their marriage in 1940. A 1949 fire at Clifton destroyed her efforts, according to the Sea Point Gazette, but I am sure Ruth was delighted with the after-effects a year or so later when everything started to grow again. In 1947, Walter bought a piece of mountainside near Botrivier. Staff of the Bolus Herbarium assisted Ruth in identification of plants she collected there. She found at least 30 varieties of protea of which some were not well known. By 1952 Ruth published a pricelist of Protea and other indigenous plants. The drawing of a *Protea longifolia* was used to illustrate her slogan 'Proteas for Your Garden!' In an article published in Veldtrust, July 1952 on Ruth's plant trade it is mentioned how she drew inspiration from Kirstenbosch. She already had stock of 2,000 plants by that time, although sales were not very good in the first year when she sold only 200 plants. She became quite well known for her seed collections. TP Stokoe collected seeds for her on his wanderings which he would then mail to her.

Ruth Middelmann drying protea seeds
(taken at her Newlands home 1982)

Weight of seeds per 1,000	
Ld argenteum	215 gr
Lsp cordifolium	110 gr
P compacta	155 gr
P cynaroides	35 gr
P eximia	45 gr
P magnifica (barbigera)	210 gr
P neriifolia	35 gr
P repens	50 gr
Source: Ruth Middelmann	

Ruth gave her first talk on the use of indigenous plants to Botanical Society in 1959. Years later Chris Barnard, the heart surgeon, commissioned her to provide indigenous plants for his garden in Bishopscourt.

In 1948, they were commissioned to supply flowers when the city of Cape Town was asked to provide specimens of proteas to London for the International Floral Exhibition in August. This followed shortly on the heels of proteas being shown at the Fleurop Conference in Paris in June.

Conservation and Organisations

In 1949, Captain EJ Scholtz gave a talk to the *Natural History Club* in which he advocated that foresters should be involved in collecting data and seeds in forest departments in the Cape Floral Kingdom and that this information should be deposited with Kirstenbosch. Kirstenbosch would then be able to check on species where there is danger of them becoming extinct, introduce such species to others areas and encourage *Botanical Society* members and the public in general to grow our wild flora. In the second part after tea, he encouraged people to grow indigenous plants in their garden and he proceeded to give some basic pointers for success. The lecture copy was sent to Prof. Compton of Kirstenbosch by Miss Edith Stevens who was then Secretary of the Cape Natural History Club and was subsequently returned to Miss Edith Stephens on 12 December 1949 together with a short note. Walter Middelmann was President of the Cape Natural History Club at the time. Seeing that Walter saved everything these documents are now available to us.

(Refer to Appendix B: Some aspects of the problem of preserving the indigenous flora of S.A.)

In 1954, Prof. HB Brian Rycroft succeeded Prof. R.H. Compton as Director of Kirstenbosch. During his term of office he was an outstanding and active propagandist for proteas, not only in South Africa, but worldwide — with talks and addresses at congresses and meetings where exhibits, flown over at his instigation, created enormous interest. Through his vision and energy, the National Botanic Gardens have become truly 'national' with a string of gardens in many of our floral regions throughout South Africa. He became a close friend of Walter and Ruth as did the next incumbent, Prof. Eloff. The ties with Kirstenbosch and its directors remained close throughout the years, until the retirement of Brian Huntley.

There is a lovely little newspaper clipping (Cape Argus 29/10/1960) about Brian Rycroft, the Director of National Botanic Gardens, Kirstenbosch on his visit to London. 'Brian had been elected as a Fellow of the Linnaean Society and following tradition, addressed his illustrious fellow Fellows on the subject of proteas. The lecture hall was decorated with a large bouquet of proteas airmailed from Kirstenbosch for the historic occasion. This giant bouquet was destined to be a gift for the Queen. The next day, Brian and his wife Maureen drove down Pall Mall to the palace to deliver the bouquet. When they left he suddenly remembered that the car had to be pushed each morning and he had visions of Maureen having to shove the car out of the palace grounds. But, fortunately it started with a royal rumble and as they drove to the gates, a policeman on duty held up his hand to stop the traffic and let them out. Suddenly the roar of London fell silent. In the stillness Brian saw the solid phalanx of hundreds of cars and several red double-decker buses immobilized for him. Then, just as suddenly, the roar returned as Brian drove into the moving stream of cars.'

Although the *Department of Nature Conservation* had already been established in 1952 under Dr. Douglas Hay, I could not find much literature in Walter's papers. Dr. Hay was appointed by the State President as Chairman and only member of a Commission of Inquiry into the future control and management of Table Mountain. Somewhere in some State Archive there should be a copy of his report.

Under Douglas Hay's leadership and to inform the public of the importance of the wild flowers around them, booklets were published, illustrated by at first Mary Maytham Kidd with later editions being supplemented by numerous botanical artists, like Cecile Omerod, Fay Anderson, Heather French, Hettie Louw and Ethel M. Dixie. The Department also published leaflets, posters and popular articles, and held various exhibitions.

It is interesting to note that it took until about 2000 for all the different organisations that had some stake or interest to agree on a single management agency. I might add that it was not an easy road as the public of Cape Town feel very strongly about how they want to see Table Mountain run!

In 1955, Dr. John Beard of Natal who later published a number of books on the proteas of Tropical Africa won the FH Ferreira memorial prize competition for his entry: *Protea, the South African National Floral Emblem*. I quote: 'The general recognition of the Protea, Suikerbossie or

Sugarbush, as our national emblem seems to have been effected more by tacit consent than by official enactment. Of the latter, we know of none, but popular references to the fact occur repeatedly. Protea mellifera (P. repens) is featured as a floral emblem on the South African coat-of-arms and Protea cynaroides appears on the reverse of the three-penny and six-penny coins. The Proteas are certainly the most strikingly beautiful of the common flowering plants in the South African flora and thus merit national recognition . . . ' More on this subject later!

The fledgling protea industry was facing trouble. The sale of plants by the Department of Nature Conservation was a problem for the new entrepreneurs who had started up nurseries to supply the public with plants. Walter Middelmann, already in 1961, via the *Flower Growers Association*, requested a meeting with the Provincial Administration regarding the matter. An audience was granted with the administrator on twenty-sixth April. The conditions of sale were that Nature Reserves, Wildflower Gardens, government departments and Provincial Administrations received preferential treatment, and that land owners in the Cape Province, whose land is 10 acres or more in extent, may purchase plants in batches of not less than 100 plants, not necessarily of the same variety. The department had stated that they were assisting and popularising wild flower plants and that they would stop if the industry could demonstrate that it could reasonably meet the demand. Discussions must have continued, because a press statement was released in only November 1962 after the latest round of discussions with the *S.A. Nurserymen's Association* (SANA) and officials of the Department of Forestry. SANA published the following:

> *For some years there has been a difference of opinion between the Department of Forestry and the S.A. Nurserymen's Association about the practice of the Departmental Tree Nurseries of supplying ornamental shrubs to the public in addition to the normal supply of forest and windbreak trees to farmers. SANA has always maintained that this practice constituted an unwarranted intrusion on the rights of private horticultural enterprise.*
>
> *Recently certain farmers' organisations approached the Department with the request that the supply of these items be discontinued. In reply the Department enunciated its policy of not entering into*

> direct competition with existing private Nurseries where these Nurseries could reasonably cater for the demand.
>
> Regarding indigenous shrubs, such as proteas and related plants, the Forestry Department feels it can give a lead in popularising this class of plant, especially in areas distant from the numerous specialist growers now registered under Provincial Authority. However, once the supply from private enterprise shows itself able to satisfy this demand (admittedly not large at present) the Department will gradually withdraw.

Apparently withdrawal took some time, because in 1963 Mr. Starke of the S.A. Nurserymen's Association wrote regarding the sale of plants:

> The original and recommendable idea of every municipality having a horticulturist on the staff is surely to provide a presentable park and trees for intelligent street planting and town landscaping. How can these functions be carried out in an efficient manner if such professional staff are required to prostitute their calling by peddling bedding and other plants to all and sundry (Farmers Weekly 6/2/1963)?

Minister Jan Haak was the first politician to decide that a protea farm would be nice to have. He had huge plans for the 1050 morgen of land he bought in Du Toit's Kloof. Fruit vineyards as well as proteas and ericas in addition to a holiday resort, with the proposed name Protea Park, was announced with great fanfare. The Cape Times of 3 November 1964 ran a humorous article on Minister Haak's desire to export proteas and his intention to practically own the whole Du Toit's Kloof pass area to grow them. The article also touched on the proposed seed selling ban:

> 'There are all sorts of horticultural bans officially operating all over the world on the dissemination of one country's seed (and plant pests) to other countries'. The article also stated: 'When it comes to seeds, nothing defeats the garden-minded aunties. In hundreds of letters from them and to them, seeds go around the world. Seeds of all sorts come and go by every mail in letters that the Post Office could not possibly get round to checking for possible contraband contents.'

Well-meaning but poorly informed people thought that if protea exports developed, a monopoly for South Africa could be created by prohibiting seed exports to potential competing countries, a measure impossible to administer.

A year later, Landbou Weekblad published a progress report on Minister Haak's 'protea show garden' and plantations. In it mention was made that he had bought the land for R50,000 from Boetie Small. Somehow, in time the whole enterprise petered out. Subsequently, the Du Toit's Kloof tunnel was dug through the property. The minister was probably handsomely rewarded for the expropriation although no records appear on this matter.

South African Wild Flower Growers Association (SAWGRA)

Until Kirstenbosch was established in 1913, few members of the public had taken much interest in the Fynbos. Another thirty to forty years were needed until the general public recognised the value of our native flora. Only then people started wanting to buy plants for their gardens and to use proteas to any extent in floristry. By the early 1960s overseas demand for protea flowers added to this interest, and commercial growing began in earnest. Firmly believing in private enterprise, Mr. Walter Middelmann got together with Frank Batchelor, who sold flowers, plants and also exported seeds, to discuss the measures put into place regarding plant sales. They felt that only an association could put their case and, to gain support and secretarial help, contacted Alan Starke, then Chairman of Starke-Ayers Ltd, about forming an organisation under the S.A. Nurserymen's Association (SANA) umbrella.

It was clear that only an organisation could deal with some of the controversial matters, such as: Should protea seed exports be prohibited to 'protect' a newly developing flower industry? A letter on the subject, addressed to the Provincial Administration, written by *Frank Batchelor* in July 1964 is worth repeating here:

We were dismayed to receive your letter, in which you prohibit the Export of Protea Seed. We feel the decision would not have been taken if the full facts had been known.

Firstly, nature herself has the most efficient safeguard for protecting our Export market of Protea blooms. Their requirements are such that the areas where they might be cultivated economically, are very limited.

Secondly the Department of Agriculture of the U.S.A. has so strict a routine fumigation on seed entering the country that the percentage of germination is fantastically low.

California where we first sold seed in 1947 and extensively travelled in 1948 seemed to be the logical area where Proteas could be grown and although my client was a commercial grower, after eighteen months every single plant had died. We revisited California in 1962 and were amazed to find that instead of any progress being made in the cultivation of Proteas, it had come to a standstill. The few plants we were proudly shown did not impress us.

The severe winter of 1962 and 1963 in Europe created a false impression of the potential market for cut blooms; buyers canvassing this country unfortunately left the impression with some people that there was an unlimited market, and irresponsible talk and articles have given impetus to the idea. The biggest weakness in the exporting of proteas is that their main flowering period coincides with peak production of flowers in Europe and to protect their growers quite a heavy duty is imposed.

We should like to mention that twenty years ago we pioneered the commercial side of Protea cultivation and marketing, and have laid out thirty eight morgen now known as 'Protea Heights' which has become world famous. We have exhibited in many cities in this country at great expense and in 1947 won a gold medal at the Horticultural show in Johannesburg. Besides this we have supplied blooms for exhibitions in Salt Lake City, Utah, Los Angeles, California, and many cities in Europe.

What appalls us is that you have taken the export of the seed business which has taken twenty years to create and placed it on the Black Market. Protea seed is so easily procurable and has so little bulk that we would say it would be impossible to control. What has been done is penalise the legitimate grower, and opened up a new avenue for the unscrupulous.

If your department is interested in developing the Export Market then we suggest funds are made available for publicity overseas and that research is speeded up. These markets could easily be denied us by having to certify blooms pest free, and finally that inspection is applied on the same basis as the fruit industry'.

Another example of the type of interchanges that took place at the time and that strengthened the need for an organisation was a complaint received from Mrs. Richfield of Bloemerf Nursery that people were offering plants too cheaply. Walter Middelmann wrote to her as follows:

The only thing we can do about it is to have a Nurserymen's Association with as many members as possible to discuss such matters and try and convince members of how foolish their actions might be. After SANA agreed to the formation of the Wild Flower Growers Group, I have not made much progress because the SANA fee is too high. But I think we can form, according to Mr. Basson, a separate but affiliated group (like the vine people) with an affiliation fee of R2.10 per head. Then, with a total fee (to run our affairs) of from R5.00 to R7.50, we could get going. Prof. Rycroft is lending us the Hall at Kirstenbosch for 11 August, 8.15 pm to have an inaugural meeting and I sincerely trust you will come and tell any other growers to do the same. Invitations will be sent out by Mr. Basson later'.

On 18 August 1965, with the support of Dr. Ruben Nel, then director of the Fruit and Food Research Institute at Stellenbosch, Walter as convener, called a meeting of interested parties, and SAWGRA was formed — affiliated to the S.A. Nurserymen's Association, which offered secretarial help.

Inaugural meeting of SAWGRA (front row: Mr. Teddy Pickstone, Mrs E van Heerden, Mr. WJ Middelmann (Chairman), Piet Carinus and Mr. AJ Basson. 2nd from left, Boetie Small. (Farmers Weekly 15 Sept. 1965

At the inaugural meeting, the first Executive of SAWGRA was elected. Walter Middelmann of Newlands was elected Chairman, and Brian Rycroft was elected President, with AJ Basson of Paarl elected Secretary. Other members were EO (Teddy) Pickstone, Mrs. E. van Heerden of Lynedoch, PJ (Piet) Carinus, JJ (Boetie) Small of Rawsonville, BW Gericke of George, F Batchelor of Stellenbosch, JC Fourie of Napier, and P Geldenhuys of Hermanus.

The report published in S.A. Nurserymen's Association publication read as follows:

> *The vital need to investigate the marketing aspects of indigenous flower growing was stressed by the Chairman, Mr WJ Middelmann, at a meeting of more than 100 wildflower growers at Stellenbosch. Some of those who attended the meeting came from as far afield as Kimberley and George. A decision was taken to form an association for the producers of indigenous flora, but its name has still to be decided.*
>
> *Mr. Middelmann was confident that the technical problems facing the export of proteas could be solved, but it was not merely a question of producing and transporting. Markets would have to be found for the flowers. 'There are no millionaires in this game' he said.*
>
> *Problems had already arisen on the local market and only an organisation such as the one now formed could do something about them. It could work in three directions; with the State, through the different publicity media, and as a forum for discussion among growers themselves on quality control, marketing and price policy. It would also keep an eye on unauthorised and backdoor growers. Since it was not a marketing board it would rely on the voluntary cooperation of its members.*
>
> *So many had now entered the wild flower field commercially without a full knowledge of the facts that serious difficulties might be expected in a few years, he said.*
>
> *A phyto-sanitary aspect or removal of the 'bugs' from the proteas so that they do not contravene plant health and quarantine restrictions, is of utmost importance. The ideal treatment has not yet been found. It is a great advance, Mr. Middelmann says, that the Fruit and Food Technology Research Institute at Stellenbosch has been charged with the task of research in the various fields mentioned. Here sound technical foundations will be laid for future marketing.*

Insistence on quality must not be forgotten, he stressed. A regular trade can be built up only on rigorous grading; selection of good blooms, long straight stems and unblemished foliage. Once a market is spoiled by poor goods and the florists are disappointed with them, the price drops and it is hard to regain what has been lost.

Whatever techniques are eventually developed, it is foolish for the South African public to indulge in land speculation on the assumption that there will be a boom in the export of proteas. They can never in their present form displace the bread and butter lines such as roses and carnations, says Mr. Middelmann. Only a few protea species may be able to cash in on the seasonal trade in flowers. It will always remain a specialised and novelty market. Flooding the market with proteas will kill the demand, just as it happened with the once-precious Strelizia in California.

Mr. Harry Wood of Stanford, said preference should be given to 'indigenous flora' rather than 'wild flowers' in the designation of the new Association. English buyers at Covent Garden and elsewhere were frightened off when one spoke of 'wild flowers'. It was decided to leave the details to the incoming committee.

The new body will be affiliated to S.A. Nurserymen's Association.

Soon there were eighty members. Prof. Brian Rycroft, Dr Marie Vogts, C. Meiring of Caledon Garden and Dr Ruben Nel were made honorary life members.

Quite early in the days of SAWGRA it was suggested that farmers should form a co-operative to pool resources and create the necessary facilities. Walter was convinced that it would be a wrong move so early in the experimental stages of export. He was convinced that it would take individuals with the willingness to accept risk to get the industry on a secure footing. In time it turned out that he was right, because a number of companies failed, whereas individuals succeeded to a reasonable extent. Hard work and dedication therein lay success.

(Refer to Appendix C: The story of SAPPEX)

Proposed SAWGRA logo drawing
Printed version on letterhead
(SAPPEX archives)

SAWGRA organised its first excursion to the weekend farm, Honingklip in Bot River, belonging to the Chairman. The invitation read: '*Members will be shown the activities of people, who from a hobby started in 1940, and on a piece of land originally acquired as a weekend retreat, have gradually built up a business in plants, seeds, fresh and dried flowers, turning an area of wild and useless mountain into a productive unit*'. With amenities being minimal at the time, only a simple tea was offered.

(Refer to Appendix D: The birth of the dried flower Industry in South Africa)

The subject of competition from the State was one of the first matters to receive attention. Through the newly formed Association, the Chairman could now, as SAWGRA, advise the Department of Forestry that the industry could now stand on its own legs. On 2 March 1966 a small article in the press announced: 'State ends its protea sales'. The Department of Forestry then offered seed to the nurserymen at price levels that could mean the ruin of the pioneer private enterprise business in this field, and once again the matter was taken to the highest level.

After requests by industry to allow harvesting of plant material on Forestry Land, the Department of Forestry started a system whereby private individuals could tender for the harvesting of certain Proteaceae on specific Forestry Land. This was subject to strict regulations regarding harvesting, access, neatness etc. This was brought to the attention of SAWGRA for comment. A private note was written on the side by Walter *'they have sent this to the neighbours concerned (Hans Visser, Jan Haak) and feel it is insufficient and we should leave it at that since the purpose is to give a chance to Boetie Small.'* Presumably, Walter meant that there was insufficient material available on that specific piece of forestry land, and that the farmers felt that they should leave it alone in order to give Boetie Small a chance to utilise this opportunity.

Meanwhile, it was reported that South Africa was one of twenty-six countries participating at the International Floralies in Paris. The stand was designed as a garden, which had a replica of the Huguenot Monument at the Cape, a gesture much appreciated by France. The garden was built up of plants and cut flowers by Mr. Bruins-Lich, Pretoria's director of parks, Mr. John Winter (then horticultural student at Kew) and Mr. Jan van Zijl of Satour. Thirty tons of dolerite rock and slasto were sent to France, followed by 18 tons of plants of 500 varieties, including a 1000-year-old Euphorbia. This was followed by regular airfreight consignments of cut flowers (including of course proteas). Subsequently, the plants were donated to the City of Paris' botanical garden. The South-African stand won the silver medal. Gold was won by Holland with England taking the bronze.

Although Fynbos Farming was not formally recognised as Agriculture, the Department was keen to support a fledgling industry. Marie Vogts, recently having moved to Betty's Bay, was charged by the Department of Agriculture to do research on proteas. After having done research on proteas as a hobby for twenty-five years, she accepted the appointment with much delight. Before seriously starting, she undertook a study tour in Europe, from Finland to Padua (Italy) where she met up with scientists she had corresponded with over a number of years, and also studied literature on the subject. Once back in the country, she started by studying proteas in their natural habitat, and commenced with observations all over the Kogelberg.

The transfer of Mrs. Marie M. Vogts, the well-known specialist on the growing of proteas, to this institute towards the end of 1964, provided a new stimulus to commercial growing of indigenous flowers. Apart from the specialised advisory service which became available to the prospective protea farmers and others, Mrs. Vogts was made available to the Department of Forestry to assist with selection of seed collection plantations in Government Forest Reserves. By this means the Department of Forestry could make available to growers and the public selected seed of Proteaceae at reasonable prices

This work introduced Mrs. Vogts to the wide diversity of genetic material, which she planted on a test plot above Betty's Bay where she was stationed. By October 1965, Dr. R.I. Nel of the Fruit and Food Technology Research Institute (FFTRI) at Stellenbosch published the first circular on '*Research and Specialised advice for the Indigenous Flower Industry*'. This circular referred to the enquiries received about indigenous plants (not just proteaceae) since 1937 when the FFTRI came into being as the Western Province Fruit Research Station with its experiment station at Bien Donné in Groot Drakenstein. It mentioned protea pioneers like Frank Batchelor of Stellenbosch and JSH Roux of Franschhoek. Somebody hit on the idea of investigating the possibility of sending Fynbos by sea freight because of the high cost of airfreight.

*Marie Vogts with SAWGRA members at
Oudebos P. cynaroides trials 1975
(Veld & Flora — The story of SAPPEX 1984)*

There is evidence in the literature that seed sales were taking off in Australia and New Zealand in the early 60s. Bernays & Bernays, solicitors in Toowoomba, Queensland were early customers of Honingklip Nurseries, having purchased seed from Ruth Middelmann as early as 1960. Two years later Bernays & Bernays reported success in raising 9 Protea, 5 Leucadendron and 2 Leucospermum species.

There is quite a bit of correspondence between Wally Stevens and Ruth. I came across a few letters dated from 1961 to 1963. These were chatty letters, not really businesslike, but more a sharing of ideas and knowledge. Wally wrote in an article for the *Journal of the Botanical Society of South Africa* in 1965 that his interest in proteaceaous plants of South Africa dated back to 1918 when he first saw a Protea mellifera (now Repens) flower in a nursery in New Zealand. He had started his own nursery in 1932. In 1963 he wrote:

We have just installed a mist propagating outfit — not because we needed it really, but it is most fascinating and of course some plants that are difficult under ordinary methods, root quite well under it. While it is standard procedure in most nurseries in New Zealand, very little is known about the uncommon plants that we specialise in. So, in a way we shall be doing some original work. Very fortunately we kept the boxes of un-germinated Leucadendrons from you last year and it is rather interesting to note that elimensis, tortum and even one platysperma (sic) have come up.

It was interesting to learn from another source, that Wally's wife, Jean Stevens was a world-renowned Iris grower.

Another snippet from a letter of Wally to Ruth: *I had a man here yesterday from South Australia. His name is Ashbey. Wants to grow all the proteas I have, referred him to you for seed. I suggested organic manure — he said that was against all the rules. Well, I don't know much about rules — here we heavily manure all our proteas with old sheep manure and they all do well.*

In 1964, Wally Stevens asked for a permit to collect specimens of plants, seeds and bulbs in South Africa. The Department of Nature Conservation in Cape Town replied that they had been inundated with requests for the collection of protected wild flowers to such an extent that they were forced to lay down a policy whereby permits would be restricted to scientific institutions undertaking approved research projects. At the time, according to Dept of Nature Conservation, there were more than three hundred registered wild flowers nurseries. Of course these nurseries not only specialised in proteas, but also in succulents, bulbs and other indigenous plants like cycads. It has never been possible to extract information from Nature Conservation as to which categories permit holders specialised in.

Duncan & Davies, probably the biggest nursery in New Zealand, regularly published a catalogue and Cultural Guide. In the 1965 edition, pages 136/7 featured 11 protea plant species for sale, each one with a short description. For instance under Protea neriifolia: *Probably the most popular Protea for New Zealand conditions, it makes a charming shrub, being easy and quick to grow and flowering on a very young bush. The cone-shaped flowers are pale salmon to deep satiny rose-pink in colour, tipped with a quaint tuft of black, velvety 'fur'. The flowers last almost three weeks when picked.*' By 1967, Duncan & Davies already had a comprehensive card system for recording several years' trials and best results for growing proteas, leucadendrons and leucospermums.

In this period, the West Australian Wildflower Society came into being and published the *W.A. Wild Flower News* quarterly. The February 1969 edition announced that Dr. J.S. Beard would give a talk on Plant Collecting in South Africa. While stationed in Natal, John had presented a preliminary account to the Annual Congress of the S.A. Association for the Advancement of Science in Nelspruit on the subject of 'The Protea species of the Summer Rainfall Area of South Africa'. John and Pam had meanwhile been transferred to Perth where he became the Director of Kings Garden. They had a lovely property on the edge of the Swan River, where I met them in 1991 during the International Protea Conference held there.

Further afield, there were already a few proteas growing at the Maui Research Station of the College of Tropical Agriculture of Hawaii, probably as a result of an exchange between institutions like Kirstenbosch and Hawaii. The first seeds were brought to Hawaii in

1964 by Dr. Sam McFadden, a visiting professor from the University of Florida, according to an article by Phil Parvin for Pacific Horticulture in 1985. The University started buying seeds from Honingklip in 1966.

Walter and Ruth Middelmann had by then already carved out a name for themselves and had embarked on numerous horticultural studies and lecture trips abroad to coincide with a visit of their clients. Invariably they would be asked to speak on cultivation of proteas and so interest grew. Very often he would write a short report of his visit for Associations he belonged to, leading to local requests for him to give talks. In 1966, in preparation for a talk on proteas to be given before the Seven Arts Club, he contacted the Argus to find out via their columns whether any members of the public knew of proteas having been grown in gardens, other than at Kirstenbosch, before 1913, and also perhaps of growers between 1913 and 1940. He wanted to know just how many people had tried to cultivate proteas before Ruth had embarked on tentative efforts to grow proteas on their mountainside property above Clifton.

Mr and Mrs WJ Middelmann, South African Protea Specialists (NZ Gardener 1967)

In 1967, having already toured Europe and the United States, Walter and Ruth Middelmann paid a visit to New Zealand. *The New Zealand Gardener*, Feb 1967 published an article on their visit and lectures and called them protea specialists. Walter was quoted to have said: 'New Zealand is by far the biggest and most consistent buyer. I believe New Zealand nurserymen grow and sell more proteas than all the nurserymen in South Africa'.

Because of the commercialisation of the wild flowers, Prof. Brian Rycroft appealed to the South-African nurserymen in 1960 to help in the conservation of our natural flora and not to exploit it for personal gain if, by so doing, the future existence of any species is likely to be threatened. He said that it was a debt that nurserymen owe to the unique South-African flora, which was providing them with the means of livelihood.

Conservation was not a new thing. Dr. Hamilton of Barberton was involved with procuring specimens of local proteas for Dr. Beard's book on 'Proteas in the Summer Rainfall Area' and came upon a small protea of recumbent growth. He convinced forestry officials not to destroy the area by planting timber. Dr. Beard later proposed that this new form of *Protea rouppelliae* be called *forma Hamiltoni* — in plain language 'Hamilton's dwarf Protea'. Although Dr. Beard stated that the dwarf could not be distinguished from *Protea rouppelliae*, the habitat of the dwarf type appeared to be firmly fixed. The area where they were found was subsequently proclaimed the Dr. Hamilton Protea Reserve.

Suddenly, proteas were receiving the attention of the daily press. For instance the Cape Times of 3 March 1962 did a comparison on California's success in cultivation of Protea and other indigenous flora with efforts at cultivation in South Africa. There were reports of proteas being successfully grown on the Scilly Islands.

Reg Weiss of the Farmer's Weekly, then Associate Editor, first wrote about the 'valuable new farming industry for the Western Cape', and interviewed Marie Vogts on her views. In time Reg became a regular reporter to be seen at agricultural events and he subsequently attended many a SAWGRA/SAPPEX meeting.

Mr. Pieter Joubert, originally from Wellington near Cape Town started a protea plant nursery in the Westonaria near Johannesburg in 1962. He claimed that he had a contact in Rome who would take ten airplane loads of flowers per day! He was in the process of establishing 150,000 plants and was making 10,000 plants at a time for sale to other growers. In 1964, Dr Joubert, who had by then become known for having the biggest protea nursery in the world, suffered irreparable damage when a grass fire, fanned by a whirlwind, destroyed eighty varieties of plants in his display area. What happened after that I have not been able to ascertain.

Some of the earliest evidence of registered Fynbos Nurseries is listed below. Some of them still exist today.

- 1952 Bergstroom Nursery, Constantia. Their printed catalogue dating back to 1965 showed not only how many different proteas were by now available for sale, but the catalogue gave a short description of the plant, and its flowering time and plant size when fully grown.
- 1955 Starke-Ayers, published a pricelist of Proteas and Ericas.
- 1959 S.R. Schreiber, 'Protea Lands', Nshongweni, Natal sold plants and seeds. He was formerly the senior horticulturist at the National Botanic Garden, Kirstenbosch
- Frank Mellish, Bergendal Kwekery, D.G. Malan Airport, Cape Town.
- Feathers Wild Flower Nurseries. Mrs. A.C. Parkes, Constantia.
- Malan Seuns Kwekery, Rosslyn
- Torfell Nursery, Richmond, Natal, Mrs. Henderson
- Victoria Wild Flower Park and Nursery, Caledon
- Ruyterplaats Nurseries, Hout Bay
- Howies
- Mountain Rose Wild Flower Nursery, Thornhill — seeds, succulents, bulbs, plants
- Transvaal Protea Nursery, Westonaria
- Leeukoppie Nursery, Cape Town — gift boxes of fresh flowers
- Golden Cape Fruits advertised protea gift boxes to the UK for R5.00 including delivery.
- Pickstone's Nurseries in Simondium, although perhaps better known for fruit trees, also was an important supplier of protea plants, with branches in Elgin and George.
- Suikerbos Kwekery, Lynn East, 1966 specialised in Proteas and Ericas.

Herbert Nash was also in the game and had a nice list of proteaceae for sale. His planting hints for proteas read as follows:

- *Proteas do well in most soils, except heavy clay. Choose a sunny, open, well-drained position.*
- *Do not prepare holes (as you would for conventional shrubs) but simply open the soil to approximately 12 inches deep, adding a little leaf compost if available — no manure or fertiliser.*

- *Immerse tin in bucket of water until all bubbles cease, drain off for about half an hour, then carefully cut down one side and across bottom of tin. The roots and soil must always remain undisturbed. Carefully transfer into the hole, cover in and firm well all round. Stones may be packed around plant to prevent accidental cultivation, because the surface soil around Proteas should never be disturbed or loosened. Leaves or straw may be placed around stem.*
- *When watering do so thoroughly, then leave till reasonably dry before repeating — never when plants are still wet. More watering will be necessary in dry summer months.*

According to an article on Babsie van Heerden, who had already successfully run her Protea Nursery near Stellenbosch since the 60s, she advocated "benign neglect" of proteas, as these plants did not need pampering. At the time it was common practice to use old jam tins to bring seedlings to maturity. Ruth Middelmann used oil tins from her husband's garage business. These were turned upside down on a sloping tray to catch the last drops of oil for use in farm implements and vehicles. Both these ladies corresponded regularly with their clients and loved to hear how their 'babies' were doing. Nurseries were very personal affairs!

Frequent references to Marie Vogts include 'Hints on growing proteas from Seeds" published by Stuttafords Garden Shop who sold seeds not only in South Africa, but also to the U.S.A. and the UK where protea plants grew successfully, although only for gardens and pot use in the Isles of Scilly, the Channel Isles and Cornwall where owners were encouraged to bring their plants indoors for the cold winters.

Dougie Ovenstone of the farm High Noon above Villiersdorp was a pioneer in eco-tourism when he developed tourist facilities and overnight amenities on his farm. Not only could visitors admire the apple orchards and related farm activities, but they were welcome to ramble in the Fynbos covered mountains. Proteas were also planted for dried flower exports to Europe. Palomino show horses, apples, Fynbos, hiking and driving routes were all there to be experienced by the visitors. High Noon closed its doors to the public in the early 2000s.

A gem is a talk by Mike Downs of Ovenstone Farm, Villiersdorp who, in a talk at the Pomological Association, stated that in 1968 he had visited an importer of flowers in Covent Garden. This company had displayed a very beautiful collection of South African Proteas in a show window in Covent Garden. He wrote:

> Within a short period of time a large crowd had gathered on the pavement outside, so large in fact that a few London Bobbies were called to disperse the masses. Unfortunately it was not the blooms which were being much admired, but the myriads of strange 'goggas' that the London sunshine had attracted out of the flower heads, and were crawling up the plate glass window. Integrated pest control means leaving the bugs in the field and 'death to all bugs' in the pack shed.

At the time even the guards of the South African Railways complained about the insects in proteas that caused great itchiness.

So here was potential for a whole new scientific investigation. Dr. J.T. Meynhardt, who was later appointed as head of the Agricultural Research Council's Horticultural Research Institute, Roodeplaat, was then a specialist Biochemistry and Plant physiologist attached to the FFTRI (Food and Fruit Technology Research Institute) at Stellenbosch. By 1965, research was already being conducted on vegetative propagation of Proteaceae by means of grafting and rooting of cuttings, particularly of hybrids. Because of the insect problem, Dr Meynhardt appointed Dick Rust as entomologist. He discovered that a small mite was responsible for the phenomenon called 'witches broom' on proteas. His recommendation was to cut and bury or burn the affected branches. Apparently Dick was also a wiz in grafting and much to the amazement of his visitors and neighbours he had a *Protea repens* in his garden with both red and white flowers. In his spare time, Dick was also a keen woodworker. He collected the wood of different protea species and made a number of small stools out of these woods, which were numbered and marked with the species used. I am lucky to have two such stools in my possession. Other recipients are John Rourke and Mr. Kachelhoffer of Indo-Atlantic for whom SAPPEX commissioned one as thanks for their support to the industry. I was also the recipient of a miniature set of minute drinking vessels on a tiny wooden tray, about the size of a coaster, with each 'cup' made out of a different protea wood.

MNR. DICK RUST, protea-kenner en -fotograaf, het 'n seldsame ronde tafeltjie uit verskeie protea-houtsoorte gemaak en dit aan dr. John Rourke, kurator van die Bolus-herbarium by die Nasionale Botaniese Tuine op Kirstenbosch geskenk. Dr. Rourke het 'n wêrelddeurbraak gemaak met sy ontdekking dat knaagdiere sommige proteasoorte bevrug. Van links verskyn dr. Rourke, mnr. Bill Storm, bekende proteakweker van Australië, mnr. Rust, en dr. Philip E. Parvin van Maut, Hawaii, 'n kenner op kommersiele kweekmetodes van proteas.

Protea wood table presented to Dr. John Rourke by Dick Rust (Landbouweekblad 24 April 1981)

Early work was also in progress on post-harvest treatment of protea flowers for shipment by sea. Results were published, with recommendations for a transit temperature of 35°F (4°C). Flowers were to be packed in plastic bags and tied down in the carton. A pulsing solution post-coldstorage was developed and proved successful with a number of protea species, Leucospermums and Serruria florida.

1965: One of the early export consignments to the UK was responsible for a considerable loss, but that was due to various reasons, namely:

1. The consignment was too small and therefore costs were too high: the freight rate on a minimum consignment of 100 kg would be lower than on the consignment which was only 62 kg.
2. The flowers arrived just before the August bank holiday weekend, at the time that trade was difficult with many florists and their customers on holiday.
3. There was a good supply of flowers in English gardens and cheap flowers were available in the markets.
4. From October onward the prices would be considerably better, when normal flowers were not so abundant.

The report, by Quick and Low, clearing and shipping agents concluded that it would be worthwhile to send over further consignments, with a minimum weight of 100 kg towards the end of October, two lots in November and possibly another one in December. For years thereafter, this was the best period for selling protea to Europe.

In 1966, Dudley D'Ewes wrote about the Cape Flora having achieved an important breakthrough in California, while reports on plantings in the Transvaal and Rhodesia were being published in the press.

Rae Steyn of Flora Cape International lived at Elephant Rock, between Kleinmond and Betty's Bay. Apart from harvesting reeds and Everlastings for the dried flower trade, she also produced commercial qualities of posies and floral portraits as well as corsages and bridal headgear of Everlastings, in addition to funeral tributes of dried protea rosettes, everlastings and leucadendron cones.

S.A. proteas at Chelsea show

Our/Cover for Jan?
20/5/69

THE ARGUS BUREAU, LONDON,
Tuesday.

SOUTH AFRICA is represented at this year's Chelsea flower show by a small but striking display of proteas and other South African blooms on the Union-Castle Line's stand in the main avenue of the grounds of the Royal Hospital.

The flowers are displayed against a background of enlarged colour photographs of many South African beauty spots and a model of the liner Windsor Castle.

I noticed, too, outstanding among a colourful array of flowers from many parts of the world on the inter-flora stand, some slender-petalled Transvaal or Barberton daisies.

The experts say that this year's show is a triumph over hazards.

WEEK-END WINDS

Besides the frosts and dull, cool weather, which damaged or retarded many plants intended for display, the week-end winds did not help those exhibiting in the open air.

In the giant marquee, however, which covers more than three acres, there is no lack of splendid colour — provided by tulips, roses, foxgloves and peonies, in an immense variety; groups of flowering trees and shrubs and rock and formal gardens constructed by some by Britain's leading designers.

In 1969, Walter and Ruth supplied proteas for the Chelsea Show in London. At that time the Union Castle Line had a stand. The flowers were displayed against a background of enlarged colour photographs of many South African beauty spots and a model of the Windsor Castle of the Union Castle Line. Subsequently participation at Chelsea was taken over first by the Dept of Foreign Affairs and subsequently by Kirstenbosch.

John Gibb Flowers Ltd of Covent Garden, early importers into the UK were so enamoured with Proteas, that their delivery vans had written on them in big flower-covered letters P r o t e a s.

SAWGRA was becoming quite concerned that the publicity given to proteas was going to encourage people to climb on the bandwagon to make a quick buck. Walter continued to warn that there were limits to the overseas market for protea cut flowers and seeds, although it did not make it any less valuable if tackled the right way. As an 'exotic' in reasonably short supply it can command a price that would be well worth the effort. But, he warned in 1965 *'there is no easy money in the protea export trade and that there are still many problems to be solved'*. At SAWGRA's symposium of 1969 Walter stated that *'proteas by the million' for export remain an illusion. Some individuals are persevering and continued flower arrangement demonstrations overseas and representation at various shows will improve the present status of the protea overseas'*.

SAWGRA sent out the following press release in December 1965:

> *Commercial Protea Growing — a warning!*
>
> *In view of many rumours and much publicity the South African Wildflower Growers (SAWGRA), through its marketing committee, feels that the time has come to warn established and intending commercial Protea growers about existing uncertainties.*
>
> *During the main Protea selling season of 1965 the local market throughout the principal centers has been oversupplied, with consequent undercutting of prices. On the auction market considerable quantities had to be destroyed at times, with consequent loss even of the railage to the growers involved. Direct supplies to agents have upset the organised market and cause a withdrawal of formerly operating minimum sales prices with a consequent 'free-for-all'. All this is aggravated by poor quality offered in many instances. Exaggerated hopes have resulted in large plantings and the advent of numerous new growers promises increased difficulties in the future.*
>
> *A potentially vast export market is said to be dependent on merely solving certain technical problems of packing, preservation, insect control etc. A number of people have since*

burnt their fingers and after realising the very unfavourable ratio of air freight weight to the value of the goods consigned, hopes are now directed to the successful outcome of experiments in shipping by sea. The State's active assistance in trying to solve technical problems over a wide front is gratefully acknowledged by SAWGRA.

However, it is imperative that the public be warned that a solution even of all technical problems will not itself ensure successful marketing on any scale. There are many factors scarcely thought of so far; opposite seasons and consequent demand, sliding scale customs duties on CIF values, highest at our best production time, embargoes, quotas, levies, imposed without warning by governments under pressure from their own agricultural interests and finally customer preference, the need to make people interested in an unknown 'exotic' flower and teaching them how to display and use it. Ensuring regular supplies of standard high class quality is an essential to interest a wider public; so far all the other factors have worked against this.

A request has been directed by SAWGRA to the State to assist with an investigation into overseas marketing possibilities. It is a wide step from such an investigation to the successful implementation of its possible findings. While SAWGRA will do everything in its power to investigate and possibly establish an overseas market, it feels that before knowing whether there really is a market, further large-scale planting for this purpose should be discouraged'

On the European market there were early pioneers too, who were willing to take a chance on a new product.

Mr. Kerscher of Agros, Blumen Import-Export, Frankfurt, requested small trial consignments of a maximum of five cartons at a time to be sent via their subsidiary in South Africa, Messrs. Springbok Flowers in Kempton Park. These cartons were to have a maximum of forty flowers per carton, pre-pulsed with 'Everbloom'. These trial consignment formed part of regular consignments of other cut flowers from the Transvaal (now Gauteng) that were sent to Agros.

Jac van Zuylen, from Rijnsburg, Holland, after receiving his first import consignment expressed concern that flowers were dead three days after arrival. He requested SAWGRA to inform him what flowers were best suited for shipment to Holland.

Another early enquiry came via the Embassy in Switzerland on behalf of Globus Stores. At the time nobody had the sort of quantities the potential buyer was interested in, and Walter as Chairman of SAWGRA, passed the enquiry to some early exporters like Frank Mellish who supplied Gift bouquets at the airport, and Mr. James Gibson who was planning to exhibit at the Ghent Flower Show in Belgium. In his reply to the Embassy, Walter suggested a cooperative venture between a few people would be the right way to deal with exports.

In South Africa Frank Batchelor, owner of Protea Heights in the Devon Valley, Stellenbosch pioneered cultivating proteas on a large portion of his 800-acre farm. He hosted numerous field days, collected seed, and allowed research to take place on his farm. During his lifetime he and his wife, Ivy, donated the farm to the then S.A. Nature Foundation, now known as World Wildlife Fund (WWF). Later, Kobus Steenkamp, a well-known personality on the protea scene, became the farm manager as well as a long-standing member of the SAWGRA and later SAPPEX Executive Committee.

One of the earliest leaflets published by SAWGRA was on 'The Successful Railage of Proteaceae Plants'. This was in connection with railing flowers from the Cape to Multiflora in Johannesburg. Multiflora was an established flower growers and farmer's marketing organization. Top growers served on the Board of Directors, including Joachim Toxopeus, Elro Braak and F. Baarnhoorn amongst others. First attempts were made to sell proteas on the auction market. Multiflora recommended that proteas should be picked not in the bud stage, but just as the bloom starts opening. This has been the ideal standard for harvesting ever since. For many years Mr. Bennie Kotze was the general manager of this organisation. He frequently came down to Cape Town for the SAWGRA and later for the SAPPEX Annual General Meetings. I used to call him 'Our man at Multiflora'.

Opening of the new complex of Multiflora flower auction, Johannesburg. Bennie Kotze, managing director.
(Farmers Weekly 7 April 1976)

An unusual request came for a marketing campaign for ARWA in 1968. The stocking company chose the protea as the theme of its advertising campaign for the new pantyhose sheer stockings. (Yes, pantyhose is a relatively recent invention!) To cater to the many different leg shapes, the firm had to do a confidential survey — length of the leg and measurement of the waist — to give factories some indications of average sizes and fittings required. The article states:

> 'In Germany women are tall, while in France we have to produce an extra small size. Italian women are extremely fashion conscious, and South African women tend to have rather large thighs. That probably explains the choice of the protea — surely the sturdiest flower of all.'

The request was for 100,000 flowers for the week before 29 March. Unfortunately there is no evidence of whether the protea was eventually used or not.

A unique feature of Cape Town has been, and still is, the flower sellers of the Grand Parade and Adderley Street in Trafalgar Place. In March

1966, these flower sellers were informed that they would, under the new Nature Conservation Ordinance which had just then been promulgated, be prohibited from picking and selling wild flowers. Only florists could obtain a permit to sell wild flowers! Naturally they left no stone unturned to reverse the decision and to allow them to obtain a permit; eventually they managed to overturn the decision and were allowed to continue their trade. For the first time Capetonians were warned that they could no longer pick flowers along the road (within 300 ft [90 m] of the centre of the road) and that it was now illegal to pick flowers from nature without a permit. Even today, so many years later, there are still people that do not know this and think that flowers are just there for the picking. There are also still people that poach on quite a large scale.

In the U.S.A., interest was continuing. Howard Asper of Escondido, California, was one of the first to publish a leaflet on 'Planting and Care for all Members of the Protea Family'. In the July 1969 issue of *Florist USA*, *Protea cynaroides* was referred to as a 'glorified artichoke' and they coined the name 'exotic' flowers.

The San Diego Floral Association asked Walter to review the book 'Proteas for Pleasure' by Sima Eliovsen, in anticipation of her visit. Sima was also due to speak at the Strybing Arboretum in Golden Gate Park. In 1963, the Arboretum had already published articles and photos of *Protea cynaroides* and made reference to Marie Vogt's book on Proteas.

Further afield interest also grew. In 1968 Walter gave a lecture at the Ebisu Public Hall, Tokyo, for Seibundo Shink-sha at Tokyo, Japan. The announcement to the members was of course in Japanese but Dr. S. Harao sent Walter a translation of the invitation card and of the travel directions that were sent out to the public. In his talk Walter said that he was sure the proteas would grow in Toba, Takamatsu, and Kyushu, if growers would go about it in the right way.

Articles on proteas were also starting to appear in florist trade magazines like *Zierplanzenbau* and *Deutsche Gaertner Börse* in Germany. There were reports of success with giant proteas and buchu bushes, as well as crassula species at the botanic gardens of Edinburgh. The *Cape Times* reported in 1962 that Chincherinchees appeared in rainbow colours in Gouda, Holland. The florists incorporated dye in the soil around the bulb and apparently the result was wonderfully successful.

In 1963, Western Australia banned the picking of wildflowers on crown land and state forests, roads, and reserves. Wildflowers could, however, be picked on private property with consent of the owner.

During this period books and articles proliferated. Some of the titles are given below, but there were no doubt very many more.

1. *Colorado State University: Agriculture and Home Economics: Dried Arrangements to enhance your home.* By Mary Grindstaff, Circular 2795
2. *California Horticultural Society*, April 1966 Report by Walter Middelmann on visit to U.S.A.
3. Dept of Commerce and Industries, *Marketing possibilities of cut flowers in the UK*, Europe and U.S.A. 1962
4. Duncan & Davies, complete catalogue and cultural guide 1966, New Zealand
5. *The Genus Protea in Tropical Africa*, by JS Beard. Jan 1963.
6. *W.A. Wild Flower News*, Newsletter of West Australian Wildflower Society. Feb. 1669 Vol 7, No. 1
7. *New Zealand Gardener*, Feb 1967 'South African seeks his native plants in New Zealand.'
8. *The Protea species of the Summer Rainfall area of South Africa* by J.S. Beard 1958
9. *Protea, the South African National Floral Emblem,* by J.S. Beard (second prize HF Ferreira Memorial Prize Competition.)

Industry Developments 1970 - 1979
South Africa

The 1970s was an interesting phase in the South African Industry. Because the idea of commercial resource utilisation was so new, all kinds of organisations, large and small, were jostling for the best position from their own point of view. It seems from the literature that there must have been many a heated debate. The industry as represented by SAWGRA was very diligent in pursuing the position of private enterprise, while reassuring authorities that it was to the advantage of conservation efforts for farmers to recognise that there was a value to this resource and that it was worthwhile looking after it and not just for short-term gain. At the other end of the scale were ill-informed nature lovers who accused farmers of 'devastation and rape of the veld'.

Internationally, sales of Australian and South African seeds were still going strong, although growers were recognising that it was better to propagate via cuttings from superior selections for uniformity of product.

There were some early breakthroughs in research by people like Van Staden, De Swardt, and Ferreira as well as Gerhard Jacobs. But, according to a lecture by Wim Tijmens of Hortis Botanicus of University of Stellenbosch in 1971, the origin of the Cape Flora remains a mystery to be solved. Apparently, in a study undertaken in Australia, pollen of proteas has been discovered as far away as Siberia.

SAWGRA

In November 1972, the executive approved the publication of a monthly newsletter for members. The first editor was Mr. Philip Brink of Elgin Proteas. The newsletter was to be published alternately in English and Afrikaans. Harry Wood of Hermanus took over as editor in 1974 (No. 5).

The first suggested regulations for ensuring the quality of South African flowers were published, particularly with reference to P. *cynaroides*. Interesting is that at that time the quality was called 'choice grade' and 'selected grade'. Cynaroides were to be packed ten per box, with flowers fully developed with a minimum stem length of 12 inches (30 cm). Already then the regulations allowed for a 15 per cent deviation (in terms of damage or crooked stems) in the case of choice grade and 5 per cent in the case of selected grade.

Department of Nature Conservation

The Department of Nature Conservation invited the Flora Conservation Committee of the National Botanic Institute at Kirstenbosch to meet with the department to formulate policies and discuss matters of mutual interest. Walter served on the Flora Conservation Committee at the time. After at first objecting that the trade had no place at these discussions, his practical experience in the field led to him being asked to comment on the proposed ordinance that wildflowers for sale must be cut and may not be broken off. He reasoned, on behalf of the Fynbos Industry, that:

> *For practical reasons, the following should be exempt from this provision:*

- All 'Everlastings' (Helichrysums, Helipterum and Phoenocoma spp.) If these were to be cut with secateurs workers would tend to cut off the whole plant; they break off neatly and easily as a rule and attempting to cut a single stem of, say Helichrysum sesamoides, would be a hopeless effort.
- All 'Reeds and Grasses' (mostly Restionaceae spp.) same reasons as above.
- Brunia spp. Fairly soft stemmed and break reasonably neatly, but this may be a doubtful item.
- The seedheads of Proteaceae for seed collection purposes: these can as a rule be broken off neatly just below the head where there is a natural weakness in their attachment. If cut, one would cut off a piece of stem unnecessarily.
- Any dried-up material, for instance dead Aloe leaves, dry Berkheya stalks etc.

About a year later, this matter was still ongoing although it had reached the draft ordinance stage, which was then given to SAWGRA for comment. Walter suggested that people who wanted to harvest on somebody else's property should not have to go into a lease with other landowners, but rather just an agreement. He set out reasons in his submission that are just as valid today, but these days there must be a formal agreement between harvesters and landowners.

Subsequently, there was regular collaboration with the Department of Nature Conservation, particularly prior to the publication in early 1975 of the new Ordinance. One of SAWGRA's aims in its constitution reads: *'To uphold the cause of Nature Conservation, where applicable, and especially in connection with optimal utilisation of floral veld as a sustained natural resource'*.

Unfortunately, it seems there is no longer formal contact between the Fynbos Industry and Nature Conservation. There is concern about the regulations that are drawn up, which are not always well thought through from a practical land use point of view.

In 1973, a meeting was held regarding protective measures for indigenous plants. Participants included Charlie Boucher, Hugh Taylor and Teddy Oliver. The meeting was chaired by Neil Fairall, who would, many years later, after retirement, become a main role player in having the Kogelberg in the Cape declared a Biosphere Reserve under the United Nations 'Man and the Biosphere' program. The idea was to draw up lists of endangered species, protected species and unprotected species, with no harvesting of endangered species allowed. The Red Data book, specifying amongst others critically endangered and vulnerable species, was subsequently published and was revised in the early 2000s by Tony Rebelo of Protea Atlas fame.

At the time, there was a quota on the import of flowers into France, with a quota of only R7,700 for dried flowers, and R21,000 for all categories of fresh flowers, which is a miserable sum on which to build export trade. SAWGRA therefore, via the Department of Commerce asked for the Agricultural Attaché in Paris to intervene. A letter came back to state that it was difficult to ask for an increased amount as the quota had not been filled in the previous year. It was, however, difficult for exporters to actively market the product and for importers to commit to importing under such restrictive quotas.

One of the most remarkable stories of the time is that of the *Orothamnus zeyheri* or 'Marsh Rose', which was thought to be extinct, until it made a re-appearance after many years and after two fire cycles. This endemic to only two small areas, one above Hermanus and the other in the core of the Kogelberg biosphere, has been playing hide and seek with botanists for sixty years and its location is now a reasonably well-kept secret except to a few. Unbelievable as it may seem, it was documented that there was a time in the early 1930/40s that this delicate bloom was harvested by informal flower sellers and was available on the streets in Cape Town. In the 70s Phillip van der Merwe successfully grafted the Marsh Rose on a wild yellow pincushion and a few plants were distributed in the hope that it could be commercialised. This unfortunately was not successful.

October 20, 1958: volunteers control alien vegetation at Kirstenbosch. Ruth Middelmann on the right.

As awareness of the importance of the Cape Floral Kingdom increased, so the effect of the destruction caused by alien invasive plants came under the loupe. Jack Marais, erstwhile farmer and later curator at Kirstenbosch National Botanic Garden, did pioneering work in eradicating Eucalyptus, bramble and other alien species from the *kloof* above Kirstenbosch during his tenure. Eventually the whole kloof was cleared. This proved very successful with a return of indigenous tree species in the clearings. As a bonus, the birdlife returned to the indigenous forest. This challenge was watched with great interest by botanists around the world.

Thanks to scientific publications, the research community could keep in touch with developments because politics had reared its ugly head. In 1975, Prof. Brian Rycroft had spent three weeks in London, with daily phone calls to Moscow trying to obtain a visa to attend a Moscow conference of the International Association of Botanic Gardens of which he was vice-president. Russian colleagues had invited him to present papers in Moscow and at the International Botanical Congress in Leningrad. As a result of being refused a visa he gave notice that he would use his influence to approach the International Council of Scientific Unions in Paris, the International Union of Biological Sciences in Norway and the International Association of Plant Taxonomists in Holland to ensure that no further scientific congresses were to be held until the host country could undertake that all invited and official guests would be allowed to attend.

The National Department of Agriculture had meanwhile put considerable efforts into research on proteas and the Agriculture Faculty of the University of Stellenbosch published a 'Management and Planning' document entitled 'The production and marketing of Proteas' on 24 April 1970. In the document they relied heavily on anecdotal information from growers, enthusiasts, Botanic gardens and of course Marie Vogts from her book 'Proteas Know them and Grow Them' which is still to this day an excellent reference book — if you can find it! Articles appeared regularly in *Farmers Weekly*, a fortnightly publication that is widely read by the farming community in South Africa and neighbouring countries.

In 1950, the value of all flower sales, including bulbs, was thought to be in the region of R1,000.000 export and R200,000 on the local market — by 1971 this had climbed to R3,500,000 exports and R800,000 on the local market. Therefore, it was not surprising that articles started appearing in the popular press with headlines like: *'Indigenous Cape flora devastated for foreign exchange'* and statements like: *'at present tons of indigenous flora, particularly Proteas and Leucadendrons are being stripped from the veld and pickers will not let up until there is nothing left'*. It was from one of such article that the dried flower industry learnt from a world-renowned authority on Leucadendrons, Dr. Ion Williams, that for the vulnerable but popular species, *Leucadendron platyspermum*, seeds must ripen in the cone. Before that farmers had tried to collect seeds for sowing into plantations with very little success. This one statement led to an innovative way of sowing: harvest the seed head on its stem just before ripening, plant the 'sticks' in the ground, and let the seed ripen and disperse by itself. Well, that made all the difference and nowadays it is not uncommon to see pale green fields where *platyspermum* has been encouraged to self-sow. So for now, the *platyspermum* is no longer 'vulnerable'.

Harry Wood, in order to promote a better understanding of utilisation and how it can go hand in hand with conservation, wrote an article entitled 'The commercial utilization of Fynbos for ornamental purposes and its influence on the natural habitat':

In 1938 sales became restricted to certain varieties not listed as 'Protected Flowers' in the Wild Flower Protection Ordinance. This was the beginning of the awareness by farmers that their Fynbos veld could produce an income from the sale of cut flowers, as, up until then, any income derived from picking by hawkers on their land was indiscriminate. Quite suddenly he now finds that the 'weeds' have become a valuable commercial product and his potential income from this source far exceeds any comparable income from grazing. There is no doubt at all that this upsurge of the commercial value of his veld has induced a new and strong awareness of its present and potential source of income. Today most farmers know the names of nearly all the commercially worthy plants and also know how to pluck the blooms without permanent damage to the plant. Conservationist and others that are aware of what is really going on are confident that the growth of the commercial value of especially Proteaceae has stimulated an incentive for their preservation and that the 'rape of the veld' is by no means as ominous as it sounds. It is interesting that of our entire indigenous flora, proteas will be the first to have received the attention of horticulturist and plant breeders for development in our own country.

Continuing bad press regarding commercialisation of proteas and greens, led to the industry's response as follows:

No objection has ever been raised by the public or by the Departments concerned when such veld is subjected to intensive grazing and thus radically altered or destroyed, nor to ploughing up, afforestation etc. These are regarded as 'normal' agricultural activities. When, however, the farmer to whom this natural growth often represents nothing but a weed, turns to plucking Proteas, Everlastings or even Reeds/ Grasses, this is regarded with horror and 'must be controlled'. While it is conceded that this sentiment is admirable, its application not only has economic repercussions for the individuals concerned, but might have the opposite effect. It is the considered opinion of those engaged in this field that precisely because through such activities the former 'weeds' have become an asset, the farmer will become a more willing collaborator in the objects of conservation than before. He owns a potentially self-sustaining resource, but needs guidance in its optimum utilisation which on his own he will only acquire by experience and time. Encouragement rather than control in the sense of limitation should rather be the aim of Nature Conservation. They would then get more farmers to preserve their mountain lands out of self-interest, to reduce overgrazing, over-burning or the spread of alien vegetation. The average farmer will never do this out of idealism, but he will follow such guidance if he can obtain more money for flowers he sells, or for lands he hires out for flower rights.

By the early 1970s turnover of protea sales, locally and overseas, had increased and interest in cultivation was increasing. People were increasingly being made aware of the value of Fynbos. Imaginative marketing ideas were starting to surface albeit on a very small scale. One of these was fresh protea flowers preserved in resin for the American market, in particular Maceys and Gimbles in New York.

In the 1970s the industry was definitely receiving attention. The S.A. Nurserymen's Association honoured the father of South Africa's commercial protea industry, Frank Batchelor with a gold medal for his pioneering work. Frank had embarked on protea growing already in 1944 at a time when there was no handbook or practical examples to guide him. He was convinced that he would find a market for the flowers. He was the first person to cultivate flowers for the trade, and much of his life was spent at improving the quality and appearance of the protea. In order the introduce them, he exhibited flowers at show in South Africa and abroad. Frank was probably best remembered for establishing an orchard of Blushing Bride (*Serruria florida*) from two plants he bought from Miss Stanford at Banhoek. The Blushing Bride, which in nature grows high in the mountains and is classified as vulnerable or near-extinction, has been saved for future generations to enjoy, provided the industry continues to grow this lovely little plant for either cut flowers or as pot plants.

In 1971, the industry named a competition after Frank Batchelor. This came about after Harry Wood suggested a prize for the best cultivated protea, to encourage others to follow Frank's example, whereupon Frank donated a floating trophy for the purpose (the whereabouts of this trophy is unknown at this stage). The committee then decided to name the competition the Batchelor Competition. It was not until 1973 that a meeting was called to create a judging committee to draw up rules and judging criteria for the Batchelor Competition. The list of people attending that meeting reads like a 'Who's who' on the Fynbos Scene at the time with names of people like Prof. Brian Rycroft, Dr. John Rourke, Dr. Marie Vogts, Mr. Gerald MacCann, Mr. (later Prof.) Gerard Jacobs.

Summary of meeting of the proposed judging committee for the Batchelor Award, 23 Nov. 1973 (SAPPEX archives)

In 1975, the first protea cultivar was announced, bred by Frank, Protea vc 'Ivy', with *P. mundii* as one parent. Frank named this cultivar for his wife. The Department of Agricultural Services determined that all cultivars should be registered with them. Pretty soon Frank had registered five cultivars, amongst which was Leucospermum cv 'Golden

Star'. I wonder what he thought when he was the first recipient of a competition named after him!

By 1972, the vegetation of the Cape Floral Kingdom was considered to be of sufficient importance that it became the subject of a series of public lectures at the University of Cape Town's Summer School. The University's Summer School is an eagerly awaited annual event at the Cape. The lectures are diverse and range over a myriad of subjects, from religion to rock & roll and have a regular and inquisitive audience.

The daily press was still picking up on the 'Millions to be made' theme, what Walter called the 'Export Myth'. Every time it was the same; 'buyers are keen to accept very much larger consignments'. People were climbing on the band wagon, without the necessary marketing being done and assured. A brand new firm with big plans on the production and sale of fresh and dried flowers arose in Villiersdorp under management of Mr. S.J. Joubert. A few years down the line, this business was bankrupt — the investment costs were too high to recoup and talk of 250,000 dried flowers being picked on ten farms and seventy ha housing thirty different species under irrigation was part of the 'myth'.

On 25 August that year, Indo-Atlantic entered into an agreement with SAWGRA as they wanted to become more involved in the developing protea industry. They made their offices available and SAWGRA would be permitted to appoint a member of their staff to act as Secretary and accountant for the Association. This staff member would be paid by Indo-Atlantic. They would bear all costs of books, stationery, printing, postage etc., and they also undertook to publish SAWGRA's quarterly newsletter and other notices and circulars. In addition they were willing to coordinate the contract for bulk carton purchases for members. This generosity meant that SAWGRA was able to utilize its income for promotions and market research. Within a short space of time, Reona Sivertsen was appointed and she did a remarkable job. The Managing Director, Mr. J.P. van Zyl was subsequently made an Honorary Member of the Association.

In September, Dr. Hay addressed SAWGRA on the objectives of the Cape Ordinance on Nature Conservation.

(Refer to Appendix E: Objectives of the Cape Ordinance on Nature Conservation, Dr. Hey)

On 7 October 1976 at a ceremonial buffet lunch at 'Protea Heights' Dr Anton Rupert on behalf of the S.A. Nature Foundation, accepted the title deeds of the property donated by Frank and Ivy Batchelor. They were given a large oil painting of themselves as a token of appreciation.

SAWGRA encouraged farmers to plant proteas in orchards for reason of better quality and control over diseases and insect infestation. It was thought desirable that the industry should now mature to being able to supply in sufficient quantities for the overseas market that was displaying increased interest. For this reason, the chairman applied to the Department of Agriculture to have protea farming recognised as an agricultural pursuit. In his motivation, Walter Middelmann submitted:

Utilisation of floral veld is an agricultural activity, conducted for commercial production, for profit, and therefore should be a matter concerning, like all other farming activities, the Department of Agricultural Technical Services, from the point of research into the goal of optimum production. The farmer using his veld for flowers, green and other decorative plant material is no different in his aim to achieve optimum production for maximum profit from another who uses his veld for grazing sheep or cattle.

It took until 1994 under Chairmanship of Maryke Middelmann, that the industry finally obtained this recognition, after being able to demonstrate that the fresh flower sector of the industry was now sufficiently mature and orchard plantation orientated.

Walter retired as active Chairman in 1976 after a twelve-year term of office and was elected Honorary Life President of the Association. The reigns were taken over by Stefaans Joubert of Villiersdorp with Barrie Gibson elected as Vice-Chairman. Stefaans, after his first term of service, brought out an 18-page Annual Report — surely a record, hopefully never to be repeated. A mere two years later in 1978, Barrie Gibson of the farm Heuningklip, near Kleinmond took over as Chairman of the Association. His father, James, had initiated early exports to the UK and now Barrie, who had some overseas working experience, was handling marketing, under the name Mountain Range Flora, while his brother Peter preferred to be the hands-on farmer.

Harry Wood of the farm Gledsmuir in Stanford, who became the Editor of the SAWGRA Newsletter 5 in 1974, held this position until 1980. The newsletter was distributed not only to members, but also to the scientific community and the popular press, who from time to time would publish interesting information to inform the larger readership. Harry was wont to write little poems about proteas. Harry was also the person who was the main author of the constitution, and whenever a change to the Constitution was brought before the Annual General Meeting, Harry would be there to ensure that things did not go wrong. Harry was the pioneer of the 'Greens' trade in heaths, leucadendrons, brunia and other veld harvested products, being the first to utilize these indigenous products at a time when others only considered harvesting proteas. These days the use of 'Cape Greens' is well entrenched in the Fynbos Industry.

In spite of early sea trials, until very recently, there was only one way to export flowers and that was by air. Flowers had been supplied by SAWGRA for a trade delegation to Austria and Germany, attended by Dr. RJ Jordan of Kirschhoff & Co (he was married to a Kirschhoff) where he complained about the shabby treatment of the flowers by the South-African Airways. 'It would appear that this matter is being pursued but that patience may be needed for the time being'. One of the biggest problems was to secure air space at short notice. This led to tentative discussions regarding sea freight, which had already been mooted in 1966 when the first trial consignment of proteas by sea was shipped to England. Only one species, namely *Protea compacta* was tested. It was very difficult to draw conclusions as there were differences in success between the Botrivier and Napier species. The only positive thing that came out of the report was that 4°C is the ideal temperature for transport and storage of protea flowers. This was adopted as the industry norm until much later when it turned out that different species require different temperatures.

Another obstacle to export efforts was the regulations relating to the grading, packing, marking and inspection of flowers for export, with different inspectors interpreting the regulations in a different way. After a lengthy and thorough submission by Kapflor (Des Walsh) regarding customer requirements, the authorities agreed to input from the industry towards changes and revisions in the regulations.

Then in 1974 with protea exports increasing, further problems arose regarding airfreight. At the time SAA had the monopoly on freight and special permission had to be granted to use other airlines. In order to encourage exports, a small subsidy was granted by the government of 4c/kg reduction on the freight rate between Cape Town and Johannesburg, but with rising airfreight costs this did not mean much. There were frequent reports of cargo being left behind in spite of freight space having been confirmed. A lengthy exchange followed between Mr. Horwood, Minister of Economic Affairs and SAWGRA, but the problem remained and airfreight space from Cape Town stayed a real problem for a very long time and was an issue taken up by the industry time and again.

Presumably trade tariff headings were promulgated by the Department of Trade and Industry quite a long time ago, but the earliest reference for tariffs that I found for the floriculture sector was in 1974:

> 06.03: Cut flowers and flower buds of a kind suitable for bouquets or for ornamental purposes, fresh, dried, dyed, bleached, impregnated or otherwise prepared.
>
> 06.04: Foliage, branches and other parts (other than flowers or buds) of trees, shrubs, bushes and other plants, and mosses, lichens and grasses, being goods of a kind suitable for bouquets or ornamental purposes, fresh, dried, dyed, bleached, impregnated or otherwise prepared.

Up in the Northern Provinces, Multiflora, a public company controlled by flower growers moved into a new marketing complex near the Johannesburg market. Modelled on the large Flower Markets in Holland and Germany, it also opened an auctioning system along the Dutch clock model. Multiflora certainly became an important destination for locally marketed floral products. One of Multiflora's directors was Rolf Jülicher an immigrant from Germany, who trained a number of people who today are still active in the flower trade; amongst them was Willem Verhoogt of Bergflora. Tiel Bluhm, who was later to raise the profile of South African proteas in Germany, also received his grounding there. Protea farmers were duly subjected to criticism from the General Manager, Mr. Bennie Kotze who complained that farmers would send their second-grade flowers instead of the top quality, thereby giving proteas a bad name.

One of South Africa's leading floriculturists was Elro Braak. Although he was not involved with proteas at all, he was amongst others a world-class flower arranger and a founder of a florist school. He was instrumental in negotiating a more favourable freight rate for flowers to allow South-African flowers to reach Europe at a reasonable price.

In 1974, the Department of Agriculture suggested that the South African Industry mouthpiece, the S.A. Wildflower Growers Association (SAWGRA) should be renamed S.A. Protea Producers and Exporters Association. The government felt that in order to embark on research as a commercial crop, the term 'wildflowers', might meet with resistance from buyers in the long term. Research would be continued not only on the Genus Protea, but also on veld utilization of proteas and related plants (the term 'Cape Greens' was coined to cover related plants). The new name was announced at the Annual General Meeting of the Association on 6 August 1974, as the *South African Protea Producers and Exporters Assocation* or *SAPPEX* for short. Earlier that day there had been an excursion to the Batchelor's farm, which was a resounding success. Stimulating talks were delivered by speakers like Mr. Brits on 'New Trends in Protea Research', Dr. Jacobs 'A Horticultural Approach' and Prof Bigalke on 'Using Nature'. The Association also hosted Phil Parvin of Hawaii, who spoke on the Hawaiian and Californian production methods and their marketing of proteas.

At a time when business was mainly a male preserve, it was refreshing to note that in the Protea Industry there were a number of ladies who played a leading role in one way or the other. We already had Rae Steyn from Keinmond and now Maxie Meyer was the first female exporter, heading Kapflor Export in Paarden Eiland, near Cape Town. When she retired, brothers Des and Bob Walsh took over. When they retired in 2008, their business was brought into the Bergflora fold.

Through the Botanical Society, close ties were forged with Kirstenbosch by the pioneers of the industry. The Compton Herbarium at Kirstenbosch was a place where landowners could find out the correct name of specimens that they identified on their own properties. In the 70s John Rourke headed the Herbarium and he embarked on a revision of Protea for the Flora of Southern Africa. Much to the chagrin of the trade, he changed the name of the popular *P. barbigera* to *P. magnifica*. Many name changes followed, some having been adopted readily, others just ignored by the trade. A formidable list was published in SAWGRA newsletter No. 7 of 1975. A protea expert, and keen hiker, John found quite a number of new specimens of Protea and other indigenous plants. He also authored books on the subject. (Dare I mention that he also acted in a short film on the history of the Cape Flora?)

Airfreight problems seemed to be permanently on the agenda. In 1976, following a discussion between Minister Heunis and Chairman Walter Middelmann, the idea surfaced that the industry should ask for a separate rate for proteas, which were far heavier than normal cut flowers, and to ask for a graduated tariff for larger consignments, i.e., instead of one rate for '250 kg plus' and ask for 250-500 kg, 500-1000 kg and 1000 kg plus. It would then be up to the exporters to club together their consignments to benefit from a lower rate. The Secretary for Commerce found that the 4c rebate was adequate, since export statistics had shown good growth in the protea industry.

The first statistics were supplied by the Department of Agricultural Economics and Marketing, Inspection Services in 1972/73. The figures covered both fresh proteas and dried flowers. More than 95 per cent of all fresh flowers and Cape Greens originated from the Cape, with small quantities being exported from the Eastern Cape and the Transvaal.

Statistics for dried flowers were expressed in cartons. In 1973, the number of cartons exported was 31,499 and this increased to 38,221 the next year. The figures for May 1976 were as follows:

Fresh flowers

Western Cape	6569 cartons	56634 kg
Eastern Cape	246 cartons	1742 kg
Transvaal	114 cartons	1150 kg

Dried flowers

Cape Town	9126 cartons	78247 kg

By 1976, there were 16 fresh flower exporter members. Statistics for December 1997 for the fresh flower trade were as follows:

Western Cape	21663 cartons	154,466 kg	R185,914,84
Eastern Cape	34 cartons	303 kg	R 325.00
Transvaal	138 cartons	1,190 kg	R 1,355,75

Later these statistics were provided to the industry by the Perishable Product Export Control Board, based on export inspection figures. The statistics were not always 100 per cent accurate and it sometimes required some prompting and cajoling to get figures on time, but we persevered and the PPECB provided a very worthwhile service to the industry.

Overseas buyers of dried flowers started to publish South African products in their catalogues, with early ones having many bad name spellings. Klaas van Zijverden was one of the early importers, followed by the firm of Jan Oudendijk. In Europe, to this day, the dried flower trade uses the name Barbigera instead of Magnifica, but worse than that, some customers got the names very wrong originally and insist on erroneously calling Compacta a Repens, with the result that we still have to check and confirm which one the customer really wants. One of the earliest dried flower customers, De Mooij, from Holland, adopted the name Pendula for Repens in 1964, a name that still sticks with one importer who got his grounding at that firm.

No protea history would be complete without mention of the Protea Ambassador, Joan Pare, a florist from Constantia who had a property, which she filled with as many species of proteaceae as she could. She not only ran a successful florist school from her premises, but she was a real globe trotter, doing many demonstrations in many countries. From a report to SAWGRA on the quality, lasting ability and arrival of flowers at one of the destination, comes this amusing snippet:

> *One of my most vivid and amusing memories of Proteas is when I flew them over for Princess Anne's wedding. They arrived three days before the wedding and the only place I had to keep them in was the bath. Each night and each morning I took them out, laid them on the floor, cleaned the bath, got in myself and when I had bathed, put them all back again. I'm sure the maids were delighted to see the last of me and I too was delighted to leave the flowers at Buckingham Palace.*

Those of us lucky enough to have seen one of her demonstrations will always remember how she could talk amusingly while deftly arranging from the back of the vase or container so that the audience could get a good view of the work in progress. She did a number of public demonstrations, for instance in the Somerset West Town Hall in 1976. She talked twenty to the dozen while making arrangements on brooms, ladders and other household items. Very amusing but also quite awe-inspiring! I remember she did something similar in the Claremont Civic Centre.

In 1976, a veld fire on Knorhoek farm of Dougie and Anne Gray laid waste to their protea plantings and pine forest. It spread to the neighbouring farm, Vergelegen, a historic farm housing an unusual monument, Camphor Trees planted there in the time of the early settlers. Fortunately, so far these magnificent trees seem to have stood the test of time.

Further up-country, proteas were known to grow, but they had hardly been surveyed. In the Blyde River Canyon a well-known 'Sugarbush', which was thought to be a *P. chodantha*, turned out to be a new species. This species was named by Lynetta Davidson, a botanist attached to Wits university, who called it *P. laetans*, meaning happiness, which is the English for the Dutch word 'Blyde'. New discoveries were still the order of the day.

The Protea, and in particular **Protea cynaroides** became the official flower of South Africa at the recommendation of the South African Society of Horticulturists. Marie Vogts sat on the election panel. It had taken a long time from the first suggestion by John Beard in 1955 for this recognition to reach fruition. The Prime Minister declared **Protea cynaroides** the National Flower in 1977. It was also declared 'the Year of the Protea'. A definitive set of protea stamps were brought out depicting different proteas. The stamps were released at Kirstenbosch on Friday 27 May 1977 by the Minister of Posts and Telecommunication, Senator J.P. van der Spuy. At the same time, a world stamp exhibition took place in Amsterdam, Holland, for which SAPPEX members supplied two tons of flowers. 20 columns of proteas, stretching from floor to ceiling in the various exhibition halls caused a sensation. A smaller, but equally effective exhibit of proteas was staged at the Kirstenbosch event, convened by Anne Gray and Reona Sivertsen. Suitably in 1977 South Africa's display at the Chelsea Show won a gold medal.

The Horticultural Research Institute's (HRI) Fynbos section offered a Farmers Day at Tygerhoek. Mother plants were established there for cross pollination to create new cultivars from which cuttings were to be made for release to the industry. Gert Brits was the person responsible and he designed a lathe house in which to establish his young cuttings. I remember that Barrie Gibson and I were summoned to Tygerhoek when the bushes were ready for taking cuttings, so that we could assist Gert in counting potential cuttings. He did not have sufficient staff to help him and he needed the numbers so that he could prepare the required cuttings bags.

Gert Brits
(Landbounuus 1990)

Gert landed himself in hot water when he recommended that growers in South Africa should also plant Australian species. Anthony Hall of the Bolus Herbarium's Threatened Plant Research Group stated: '... *it would seem foolhardy to bring more Australian plants here and to risk further invasions of the Fynbos'*. Gert always wrote extensive reports on his trips to other countries and shared his knowledge with many researchers and growers wherever he went. In addition he wrote numerous articles for *Farmers Weekly* and *Landbouweekblad*. In 1990, Gert was recognised as the Agriculturist of the Year by the Agricultural Writers' Association of Western Cape.

When Gert retired from the ARC (having been offered early retirement and a nice package because the ARC needed to restructure) he took up growing proteas as pot plants on contract to the nursery trade.

He was honoured at the IPA Conference held in Cape Town in 1998 for the exceptional work he had done on proteas, particularly on hybridization; work that benefited protea cultivation worldwide.

Sharon von Broembsen (Landbounuus 3 Oct 1986)

Sharon von Broemsen of the Department of Agriculture initiated research on diseases of proteas and gave a name to all those dreaded things that infected proteas, splotches, discolouration, die-back, and suddenly there was a name for all these nasties.

In 1978, fires raged on the slopes of the Olifants river mountains, moving towards Citrusdal. Fortunately, no fruit orchards were damaged, but damage to the flower resource ran into several hundred thousand Rand, according to Denis Shaw. It was one of the worst veld fires they had in the area for a number of years.

Under the leadership of the then Dr. Gerhard Jacobs of the University of Stellenbosch, many trials on protea growing and disease prevention took place at Frank Batchelor's farm 'Protea Heights'. In order to have Leucospermum blooms during the overseas marketing window, Dr. Jacobs commenced on disbudding trials to delay flowering time, as reported in *Landbouweekblad* of 1978. That followed on earlier trials with daylight lengthening through overhead lights.

In 1978, protea growers donated the first protea selections to the Department for release to the industry. 'Flamespike'. 'Pink Star' and 'Firedance' were donated by the S.A. Nature Foundation (now WWF). One of these was a cross between *L. cordifolium* and *L. tottum*. 'Yellow Bird' came from the Middelmann family of Honingklip near Botrivier. 'Caroline' was donated by Mrs. P. Barlow of Rustenburg, Stellenbosch.

Shortly after Barrie Gibson took over the reigns as Chairman of SAPPEX in 1978 the exporters got together and financed a brochure featuring Proteas and Cape Greens with details regarding flowering months. Wording was in English, Dutch and German.

First SAPPEX fresh flower brochure 1979

Petrus Roux addressing members at the Open Day on his farm, Elandsrivier 1978

Two farmers' days were held in 1978, one in March in Riviersonderend, and a big event later in the year in Villiersdorp at the farm of Petrus Roux. It was one of the first SAPPEX events that I attended and I will never forget the sight of apple bins loaded with people being pulled up the mountain by 6 tractors so that we could view the proteas growing in between the fruit trees and higher against the mountain. If I recollect correctly, he had planted Blushing Bride between the apple trees. Such was the importance of the fledgling industry that the Minister of Agriculture joined the meeting. During the morning there was a flower arranging demonstration by Joan Pare, while Honingklip Dryflowers had an exhibit of dried flowers and grasses suitable for export.

SAPPEX was granted financial assistance from the Department of Trade to participate in trade shows in Britain, France, Germany and The Netherlands. The assistance was for freight, accommodation and sustenance. Years later such assistance would include cost of the stand.

A breakthrough was achieved by Dr. Gerard Jacobs at Protea Heights when research results showed that by disbudding flowering of pincushion could be delayed so that the plants would come into flower considerably later. This would assist farmers to have marketable blooms at the time when proteas were required in Europe.

SAPPEX in an indirect way, via Walter Middelmann became involved in Nature Conservation in respect of Law Enforcement problems. The Botanical Society's Flora Conservation was invited and Brian Rycroft thought it would be a good idea for Walter to attend in view of his intimate and wide knowledge of the trade, and as a Botanical Society representative. A number of legislation issues were discussed (also regarding trade in other flora like cycads and succulents). Walter proposed that the registration of nurseries per se should be done away with as the majority were just 'flora growers'. It should just be stated on the registration whether the grower was also a nursery or not. Many of the issues discussed at this meeting are just as topical today and it is a pity that the regular meetings of the trade and Nature Conservation, together with the Flora Conservation Committee of the Botanical Society, no longer take place.

International Developments

Until the formation of their own associations the Australian, Hawaiian, Californian and New Zealand scientists relied on information coming out of South Africa via organisations like the Plant Propagators Society and the International Society for Horticultural Science. Phil Parvin of Hawaii was considered the foremost scientist, while Walter and Ruth Middelmann and other seed exporters supplied buyers with leaflets on how to grow proteas. The Botanical Society at Kirstenbosch distributed excess seed to their members all around the world and interest continued to grow in other regions. SAPPEX news also filtered to other parts of the world. Some interesting snippets of information were found regarding protea growing elsewhere.

England

Meanwhile, in England, seed and bulb dealer M. Holtzhausen of Cornwall published a catalogue, which included seeds of Leucadendrons, Heaths and Proteas, including Blushing Bride and Silver Tree. Presumably these were distributed to the most southern parts of the UK and the Channel Islands.

Holland

Klaas van Zijverden quite early on published a promotional leaflet, in full colour, portraying an arrangement mainly with Cynaroides and Berzelia.

Corsica

This island is not mentioned often as far as protea development is concerned, and I have come across only one reference, headed 'Be warned South African Wildflower Propagators!!' dated 1973 by W.P. Burger, Agricultural Attaché in Paris, who warned that our winter flowering species would grow quite nicely in the Mediterranean and that it would bloom at a time that coincides with the marketing window in Europe. The danger of course was that their proximity to the market would land the blooms cheaper than from South Africa. No mention was made of the high input costs of production in Europe. It was accompanied by a picture of a small protea plantation on Corsica. He wrote that he had stumbled across the proteas by accident when on a visit to the Corsican Agricultural Research station at San Guiliano while viewing some commercial citrus orchards. In the accompanying letter he wrote, '*Consider this a warning for South African protea growers to pool their resources and make a serious effort to supply the tremendous demand for proteas in Europe before the Europeans discover that they can do so themselves*'.

Guatemala

In 1975, in a letter from Phil Parvin to Walter Middelmann, mention is made of protea being sold from Guatemala, a fact he learnt from a market report from San Francisco.

Germany

In 1971, in Middelmann family correspondence is a newspaper cutting sent by one of their cousins, which mentioned that *Protea cynaroides* bloomed for the very first time in the botanic garden in München.

Australia

There is correspondence between Walter Middelmann and Roger Hall of the firm C.F. Newman & Son dated 1970 regarding an old catalogue featuring protea. *'Our only old C.F. Newman & Son catalogue is dated 1894/5. I've got photostat copies of only 2 pages with reference to plants indigenous to South Africa'*. Walter had mentioned this to people at Kirstenbosch and to some nursery colleagues who were amazed that proteas were already mentioned in a catalogue in the 1800s. This raised the question as to where the seed would have come from. Certainly at the time nobody in South Africa had such material for sale, neither plants nor seed, but somebody must have collected the seeds and taken them to Australia.

Peter Matthews was a leading nurseryman in the Dandenongs outside Melbourne. Definitely being one of the pioneers in Australia, he entered into correspondence with Harry Wood in South Africa regarding cuttings material of proteas that would be of value in the cut-flower industry in Australia. At the time, Proteaflora was growing 10,000 cuttings per year with a success rate of 50 per cent. He wrote: 'There are no nurserymen other than myself doing this kind of thing in Australia at the moment, and I have no one with whom I can compare notes.' Until quite recently there was no nursery in South Africa that could compete with the Matthews family's nursery. Peter led the formation of the Protea Association of Australia, which was founded in 1978. Shortly after its inauguration there were thirty members in this association.

U.S.A.

The Protea Growers numbered twenty four in California, mainly in the Escondido Valley where protea plantations were seen to be rising between plantations of Avocado trees. The hills and valleys of Escondido presented a difficult terrain, but this was no obstacle to the pioneers as they were quite used to farming there. Barbara and Dick LaRue, Barbara and Ray Schatz, Howard Asper Jr and Fred Mayer became well-known names in the industry. Sea-Bar Nursery (LaRue) produced a lovely fold-open brochure in full colour — probably the first one! They offered this for sale to other growers who could put their own name and logo on the back. Howard Asper a chiropractor by profession, who, after retiring at 50, decided to go full time into growing, having started in his student years with cymbidium orchids, and later rare camellias. Finally, he settled on growing proteas after attending lectures and slide shows at the Arboretum in Los Angeles. By 1974, he had been at it 12 years and had 10,000 plants on 38 acres at 1000 ft frost-free elevation in the Escondido Valley.

It was at Fred Meyer that I first saw a *Protea cynaroides* with very convoluted long stems and I couldn't help wondering if perhaps they received too much nourishment. It was also my very first overnight stay in the area where Robert and I kept each other awake by tossing and turning on our one and only experience on a water bed; what a wobbly affair. I can think of only one good use for such a bed — putting babies to sleep!

The Protea Growers Association of California was formed in the 1970s with Fred Meyer as secretary and Bill Clurter as chairman. Of the good ideas that came out of there, copied in other countries, was the 'flower contest' and attempts at standardisation of product. There was a proposal to support scientific studies on protea diseases, with some sort of levy: 'Considering what I have invested in land, plants, irrigation system etc, the amount required to support research appears to be rather paltry' Amusing, but wise, was the following statement by the editor of their newsletter: *'I look upon it as the best insurance buy available to a protea grower. Of course I would never use this Newsletter as a means of influencing other members, but if I were that sort of person, I'd make a strong statement supporting Dr. Cho's proposal'*.

In the magazine *Flower and Garden*, June 1970 (U.S.A.) an article starts like this:*'Flowers from Mars could be no stranger than the exotic blooms called proteas now being sold in florists' shops around the country. Their weird and wonderful form, vivid colours and long-lasting qualities have endeared them to florists who pay up to $12.00 a dozen for them wholesale. A single bloom may bring $3 in New York, but it would take only two or three of them, combined with something less expensive, to make a most dramatic arrangement. The proteaceae family are native only to the southern hemisphere — that is, South Africa, Chile and Australia.'*

I always wondered why the Arboretum of the University of California was so extensive. According to an article in Pacific Horticulture, the

Arboretum at Irvine had its origins in 1964 as a nursery and holding area for plants destined for landscaping the university campus. The staff of the school of Biological Sciences used some of the land for hybridisation of daffodils, and it soon became an experimental garden. At the time, the arboretum was concerned primarily with conservation and preservation of endangered plants. The scope was broadened to provide an educational and recreational resource for the surrounding community. In 1968-69, Dean McHendry and Dr. Ray Collett were the people who started a protea collection at the University of California's Santa Cruz campus. In the 1970s they expanded their Australian section and this interesting snippet was published: *'Dr. Collett and staff have completed the building of a high mound into which heavy boulders have been embedded. The sandy soil mix and the height provide the drainage required by many plants from 'down under'. The rocks, particularly on the south side, store heat that protects sensitive plants during cold nights!'*

HortScience (America) Vol 8(4) August 1973 published a special issue on proteas, with a scientific article on 'Developmental Research for a New Cut Flower Crop' by Philip E Parvin and Richard A Criley and Richard M Bullock of University of Hawaii, Honolulu. Phil Parvin lectured on proteas at the International Plant Propagators Society in San Diego in 1974 and so started his long association with proteas. Phil Parvin became a frequent visitor to many protea-producing countries and a friend of many. In 1975, through the Society for Horticultural Science, which brought him into contact with Dr. J.T. 'Kleintjie' Meynhardt in South Africa, he undertook an extended trip around the world to make contact with nurserymen, growers, and scientists, and to collect plant material for vegetative propagation in Hawaii and seed sources for development of Australian species. His later involvement with the International Protea Association is legendary.

Phil continued to write numerous popular articles that appeared in publications such as 'American Farmer' and penned innumerable scientific articles. Certainly his name appeared whenever Proteas of Hawaii were mentioned. The Kula research station on Maui was established from seed stock in 1964 and now they produce the most wonderful 'sunbursts', the name they have given *Leucospermums*. Pride of place in their experimental garden was, according to a report in 1976, a *Protea aristata*. Apparently this species was thought to be extinct in South Africa until a botany class found a single thriving plant. They returned with a tip cutting that thrived and eventually a seed from that plant found its way to Kula where this almost lost protea was reported to be in full bloom. In the U.S.A., growers thought up the most interesting names for commercialisation of the protea, with names like Pink Mink (for *Protea neriifolia*) Duchess, Princess etc.

In 1977, Walter again visited Hawaii where he and Ruth were met by Phil Parvin who presented both of them with a lei of cordifolium. They visited the Protea Research Station at Kula and were impressed by their work as can be seen from Walter's subsequent report: *Plots have been laid out. The creation of hybrids by hand pollination is being worked on and selection of good forms and consequent vegetative propagation is pursued. Part of their material comes from nurseries in California, mostly from cuttings. A gene pool is being built up, from seedlings of a maximum number of different species of the various South African proteaceae, but also of Banksias, Dryandra and other Australian members of the family. Even meristem culture is already being thought of, and everything is being done with the aim of commercial production for export to the U.S. mainland as well as to Japan. Of great importance therefore is not only research into growing methods, but also into post-harvest treatment such as pre-cooling, pulsing, packing and what happens during transport. The fact that all this is coordinated into one single Institution, situated in the main production area, is in my opinion of considerable advantage compared with us here (RSA) where the direction of research lies 1500 km away from where the growers are and where post-harvest considerations have scarcely been taken into account despite the fact that with the volume of exports now reached, this should be of the highest priority.'*

The Maui Sun published a remarkable history of the Hawaiian Industry in 1977 with a lovely line drawing by Garoscia. In this article, mention is made of Gordon Doty, one of the early growers, who first planted protea in 1972 and had his first harvest in 1975. Dannie Rhea and Carver Wilson were still students when they decided to grow proteas and during their final university exams started long phone calls and correspondence with Phil. Before long they established a protea farm.

A Protea Industry analysis of Hawaii was done by the end 1978 listing the current state of knowledge, which in its conclusion identified and prioritised bottlenecks. This was a report to the Agricultural Coordinating

Committee and was probably done to convince the Department of Agriculture of the necessity of investing time and money into the protea project on the islands.

Japan

In Japan, a culture change was taking place in the floral industry. Here, the floral arrangements traditionally change to coincide with seasons and holidays. The young women in Japan were found to have less and less time for flower arranging and would use florists for their requirements instead of going to the market themselves. In addition, the fact that modern Japanese homes have Western and Japanese rooms requires that each have the correct type of arrangement. A recent development is to use flowers as gifts. As customs changed and Western influence spread, the use of flowers in a non-traditional sense became more widespread.

The Consular offices and trade attaches were very active in promoting South African products and would arrange for trade missions, shows and displays. South-African proteas were featured in a wedding bouquet show at a Tokyo Hotel, amongst others. Thus, interest was stimulated around the world.

In 1977, Fukukaen Nursery visited South Africa to learn more about proteas. They had already bought small lots of Proteas in Hawaii for their flower auctions in Nagoya. They produced a colour brochure featuring *P cynaroides*, *P repens* as well as gladioli and watsonia that they grew from South-African bulbs.

Israel

An American, Nathaniel Hess, was such a frequent visitor to Israel that he and his wife bought a flat in Jerusalem. He made a second career out of two hobbies, namely landscaping and horticulture. He asked Howard Asper of California to survey Israel for suitable sites to grow proteaceae, paid for investigations by the Volcani Institute and went into 50:50 partnerships with a number of growers.

The Volcani Centre in Israel played a pivotal role in protea-growing efforts in Israel. Under the leadership of Jaacov Ben-Jaacov seeds were imported and trials were done to see which species would do well under their difficult circumstances. Israel was eventually to do lots of grafting on alkaline-resistant protea rootstocks. The biggest success was Safari Sunset, which they grew on a very large scale, harvested mechanically and shipped to Europe in huge crates. The Israel quality was, however, not as good as those coming from South Africa and New Zealand. This cultivar was bred from *L salignum* x *L laureolum* by Wallace Stevens in Wanganui, New Zealand, from seeds supplied by Miss Marie Stegmans of South Africa. A lot of the breeding work was done for Wally Stevens by his son-in-law Ian Bell. Safari Sunset is still the only proteaceae that is categorised in the flower records of the Dutch flower auctions.

Another well-known personality from Israel is Prof. Abraham Halevy of the Hebrew University of Jerusalem's department of Horticulture, who through regular attendance at IPA Conferences showed the contribution his department made to protea culture.

New Zealand

Mr. John Salinger who was with the Horticultural Division, Department of Agriculture, later joined Massey University as a Senior Lecturer in Horticulture. He became the leading light in promoting proteas as a horticultural crop and wrote numerous popular articles on the subject. Joy Amos, the Horticultural Advisory Officer, acted as Secretary to the Protea Growers of New Zealand and Peter Dow & Company was one of the nurseries selling proteaceae plants. An early member was Jack Harré who later authored some interesting publications on proteaceae. The different associations around the world started to exchange newsletters with each other and so increased knowledge and stimulated interaction between growers around the globe.

A paper was published on 'The Evolution of the Proteaceae and Cultural Implications' by PJ Hocking and MB Thomas of Lincoln College and it is of interest to quote from the conclusion: '*The Proteaceae are an extremely old Angiosperm family which probably arose in the distant past in the tropical regions from where they became widespread. They migrated south and some adapted to the different climates encountered. Early in their development an ancestral line became adapted to infertile soils by developing proteoid roots and physiological tolerance of low nutrient levels*'.

Madeira

Florialis, a subsidiary of Blandy Brothers Co Lta, were the first to grow proteas on Madeira in Funchal in 1970 and offered South African proteas and leucospermums for sale. They also grew Waratah from Australia.

The International Protea Association (IPA)

An International Horticultural Society Congress was held in Sydney, Australia, in August 1978. Dr. Jacob Ben-Jaacov of Israel presented a poster on 'Commercial Production of Protea in Israel', while South Africa (Gert Brits) presented a poster 'Proteas — a newly developing world ornamental crop'. During this Congress a couple of protea people got together at the suggestion of Peter Mathews of Australia who had galvanized Dr. Phil Parvin to send around a memorandum to people he knew around the world:

> "Aloha Friends and colleagues, 21 July 1978
>
> *Peter Mathews, Proteaflora, Silvan, Australia, has been instrumental in communicating with many of us regarding the possible mutual benefits of some sort of international dialogue on protea marketing. As Peter points out, the International Horticultural Congress in Sydney offers a unique opportunity for people interested in protea production from South Africa, Israel, Australia, New Zealand, California and Hawaii to get together to chat. The suggestion has been made that Thursday evening, August 17 at 8:00 p.m. would be appropriate. Unless we hear other suggestions, let us assume that the venue will be the Wentworth Hotel. I will secure a meeting room when I check in on 14 August. Please save that time, and I hope you will meet with us to talk Proteas".*

Phil Parvin presided and Peter Mathews outlined the purpose which he stated to be:

To explore the possibility of a loose affiliation of Protea-growing countries for the purpose of:

1. Cooperating in the worldwide marketing process in such matters as
   a) the use of agreed marketing names for the various species
   b) the way in which differing flowering times of species from one country to another can assist in a continuity of supply
   c) common standards for flower marketing

2. Sharing information in regard to such matters as
   a) The production of new cultivars
   b) Packing and fumigation procedures
   c) Improvements in cultivation procedures
   d) Acreage of plantings, and proposed plantings
   e) A schedule of flowering times of various species

Newsletters were considered as a medium of exchange of information, to be edited by Peter Mathews and distributed to everyone. South Africa had already been appointed as the world Registrar of varieties and cultivars. Gert Brits of South Africa would take responsibility for the Protea Register.

From the twenty-two people who attended, a steering committee was elected. They were:

Dr. Phil Parvin	(Hawaii)
Peter Mathews	(Australia)
Ray Schatz	(California)
Jacob Ben-Jaacov	(Israel)
Dr Brian Rycroft	(South Africa)
Jack Harré	(New Zealand)

Dr. Rycroft, representing Kirstenbosch National Botanic Gardens, was requested to make contact with SAPPEX and ask for representation from that organisation.

Resulting from this meeting, the first International Protea Conference was held from 4-8 October 1981 in Kallista, Victoria, Australia, where the proposed constitution of the newly formed association was tabled for approval. Walter attended on behalf of SAPPEX and became the representative for South Africa as well as a founder member of IPA. A highlight was the area report from each country, a trend that is being continued to this day. It is not recorded who suggested that an IPA Conference should take place in a different protea-growing country every two years, but adopting this system certainly has meant that growers and scientists were given the opportunity to see how things are done around the globe. As an added benefit, it also gave local growers who were not able to travel, the opportunity to meet international leaders in the field of protea growing, marketing and research, as well as meeting fellow growers from other parts of the globe.

Given below is a list of the conferences that have been conducted, almost every two years, starting from 1981.

1981 Australia, Melbourne
1983 U.S.A., Hawaii, Seattle, Southern California
1985 South Africa, Johannesburg, Cape Town
1987 New Zealand, Auckland
1989 U.S.A., San Diego
1991 Australia, Perth
1993 Zimbabwe, Harare
1996 Israel, Jerusalem
1998 South Africa, Cape Town
2000 Tenerife, Santa Cruz
2002 Hawaii, Maui
2004 Australia, Melbourne
2006 U.S.A., San Diego
2008 South Africa, Stellenbosch
2010 Portugal, Lisbon (as part of the ISHS Congress)

Industry Developments 1980 - 1985
South Africa

This was a difficult period for South Africa. European sanctions against South-African products had already started to affect the canning and wine industries, and eventually this spilled over to the flower industry, with 1980 showing negative growth for the first time. Flower poaching made headlines.

For the second time in its history, a series of protea lectures were given at the University of Cape Town (UCT) from 29 September to 13 November in 1980. The course was coordinated by Dr. Byron Lamont, a visiting lecturer in Botany from Western Australia, with a team of experts such as Marie Vogts, Gert Brits, Sharon von Broembsen, John Winter and Gerard Jacobs.

Then a report appeared in the press that harvesting was taking place where it should not be. Large-scale picking of 'Prince of Wales' Erica had been noticed in the Kleinmond Nature Reserve. There were conflicting reports on who was responsible. There was mudslinging and finger pointing in various directions. Of course this type of publicity did not enhance the image of the industry. In the end the Department of Nature Conservation banned harvesting in the reserve and stated: '*As far as the principle of the wild flower trade is concerned, it must be pointed out that the wise use of renewable natural resources is an accepted form of conservation management. The problems start when overuse, either through ignorance or greed, affects the stability of natural populations.*'

Both conservationist and the trade had very good arguments for and against harvesting, and this debate continues to this day. Industry involvement with conservation bodies has done much to dispel the facts from the myths, but even so, the industry remains concerned at over harvesting in some areas, and unauthorised poaching across fences. Both the fanatics on the one hand and unscrupulous pickers on the other hand are of course wrong! Fines from the Department of Nature Conservation were ludicrously low and were no impediment.

This letter, written in 1981 under the name 'Everlastings' is a good example of the debate:

> I farm in the Gansbaai/Stanford district. Everlastings have always been there and our labourers and their ancestors must have picked the flower heads and made bundles for over a hundred years. This has provided them and the farm with some extra income. The plants re-grow every year until they just get too old. Then the time is ripe to burn the mountain slopes and the cycle can start again. This is a natural source of beauty which is of benefit to man. A journalist writes in the Farmers Weekly of 25 March 1981 that the buyer of an overseas firm says: 'they have been picked to extinction'. I can only reply to this that he doesn't know what he is talking about. It is impossible to eradicate this plant unless you pull the whole plant out of the ground, roots and all, not an easy task! It is true, Everlastings have become scarce, but for other reasons: farmers plough them under the ground to plant more profitable crops, and they are overrun by alien vegetation i.e. hakea and by pine trees! It is the way the land is utilised and mismanaged which is causing the everlastings to become scarce, not the picking of it!

South Africa continued to show Cape Flora overseas and won a gold medal at the 1980 Southport Show where the local newspaper ran a banner headline 'Blushing Bride steals the show'. At the 1981 Chelsea Show, South Africa's two-ton display received the overseas exhibitors trophy, the Wilkinson Sword, as well as a gold medal. The Queen who had visited the stand called it 'breathtaking'. The display was built up on a 10m x 13m trellis, partly made of driftwood and was assembled by Ronelle Henning of the S.A. Embassy and Pam Simcock, one of Britain's foremost floral designers. Since that time the South-African exhibit has regularly won gold as well as the coveted Wilkinson Sword. Pam was also the designer who made the South-African arrangements in Westminster Abbey for the Queen's silver jubilee celebrations where sixty-four countries were invited to fill the cathedral with floral exhibits. In those days, David Davidson designed and built up the ever-popular South-African stand, which was regularly inspected by the Queen and her entourage. South-African flowers were also highly acclaimed at the Winchester Cathedral flower festival. In spite of these high-profile events and much interest from the public, sales to the UK was still slow and sporadic. It is of course highly likely that a lot of proteas entered the UK via the auctions in Aalsmeer in Holland, or via the specialist importers.

The buyers in Europe, when farmers started to cultivate proteas, became more selective, depending of course on whether there was a shortage or not! In times of plentiful supply, the buyers would complain about the quality, but come peak times they were willing to take what they could, bent stems, bug-eaten leaves and all. In particular, the *Protea barbigera* (now called *P magnifica*), which flowers at the time when Europe is looking for flowers, was very popular. It is very strange that the Porterville variant opens completely, while the Botrivier variant remains semi-closed, although such differences can also be seen between *P cynaroides*. At the time, numerous producers did not put their flowers in water after harvesting, but it was recommended that flowers should be harvested before the heat of the day, and then put immediately in buckets of clean water. Information about new developments were all good and well for the producers who became members of SAPPEX, but others refused as they felt they could manage without having to pay membership fees, and in all probability heard about new developments via the grapevine. I had a nice word for these people at a talk I gave to Agricultural Writers Association; I called them 'rug-ryers' 'riding on the back of others'.

And then the bombshell! TRAFFIC (Trade Records Analysis of Flora and Fauna in commerce) International Bulletin (funded by the people's Trust for Endangered Species) published in Vol 11, Nos 9 & 10 in December 1980, an article entitled '*South Africa — an entire Plant Kingdom threatened with extinction*' Although many of the facts about the Cape Floral Kingdom were correct, the following text earned them the ire of the industry:

It is impossible to escape from the impression that the Kingdom is being over-exploited for the financial and political benefit of the ruling regime. The rural Coloured population of the Cape are sent out into the mountains to gather flowers in return for subsistence wages. Almost invariably in cases where arrests for illegal thieving are made, it is the illiterate Coloured pickers who are taken to Court, while their White bosses remain free on the plea that the thieving was done without their authority. The present massive destruction for personal or official profit of the Cape Floristic Kingdom is a cause for international concern, as an extremely valuable part of the Earth's floral heritage is threatened with destruction. A massive campaign is needed, particularly in West Germany and the Netherlands to discourage the purchase of South African wild flowers.

Walter Middelmann drafted a response to all this and called it a *'scandalous, even libelous piece of distorted reporting and emotional propaganda with a strong political twist trying to play up to anti-South African agitation abroad"* He countered the arguments with statements like: *'All of us who are active in this field, on the contrary, work to the purpose of creating a sustained yield resource and thus act towards maintaining such veld which to landowners formerly was valueless and thus left to destructive forces such as invasive plants, overgrazing, over-burning etc. Wise use of the veld is an incentive for its preservation; the fact that this means money to the owner is its greatest safeguard. Far from the Coloured farm population being 'sent out to the mountains to gather the flowers in return for subsistence wages' a considerable number of such farm labourers, especially in some of the formerly poorest areas, have been given an opportunity for regular employment at rates unheard of in these areas previously. New employment possibilities have been created on quite a wide front. The slur implied in the remarks about the Coloured pickers being taken to court while the White bosses remain free, is clearly playing up to the prejudices of an overseas public which has been subjected to propaganda of a similar kind over the years in other fields and is ready to believe only the worst about our country. So are the words about the 'ruling regime' and its officials. One can only hope that knowledgeable people will not fall for this drivel, clothed as it is in quite factual statements about the Cape Floral Kingdom as such.*

(Refer to Appendix F: Survey Wild Flower Production)

The Botanical Society's Flora Conservation Committee sent a strongly worded letter to Traffic refuting the charges. Unfortunately it was not the end of the matter, as letters to Newspapers appeared discouraging people from accepted South-African products. SAPPEX encouraged producers and exporters to change the 'Afrikaans' names to make the task of the importers abroad as smooth as possible. The name changes needed to be made to remove any reference to South Africa. The cartons should not state any word with a South-African connotation, like 'safari' Cape Mix' or 'Protea', should not contain newspaper as packaging material, and no Koki pen markings.

In the same year, *Protea odorata* went extinct in nature after a veld fire in Mamre, which had probably been lit to improve grazing. Those that had not been burnt were subsequently ploughed under. Apparently the land, which belonged to the S.A. Railways was leased to a local farmer and the Mamre mission station. A few seeds were saved and were put into the Bolus Herbarium for ex-situ propagation.

There were moves afoot to build a huge dam on the Palmiet River above Kleinmond. This was met with strenuous opposition from the scientific community. They found that the dam would only meet Cape Town's needs for five years, and in the process would destroy that last stretch of Cape lowland riverine forests, and some rare protea species, including the extremely rare marsh rose. A vast area would flood one of the last strongholds of unspoilt mountain Fynbos. As a result of a joint investigation, a much smaller pump storage station was eventually built much higher up, and other alternatives for the long-term needs of Cape Town were investigated. It had a very positive outcome because the publicity had served to mobilise a group of people to lobby for preservation of the 'Kogelberg' as a biosphere reserve.

In 1980, Frank Bathelor's wife, Ivy died aged 81. The Cape Times referred to her as one of the pioneers of South Africa's protea industry. Another well-known personality who died that year was Dudley d'Ewes, who wrote a regular gardening column for the Cape Times, and often quoted from SAPPEX newsletters when writing on Fynbos conservation issues. He was a stalwart of the Betty's Bay hack group (removing alien plants from the environment).

Ongoing problems with airfreight from Cape Town, particularly in November and December, combined with temperature fluctuations at Johannesburg and Nairobi (those were the days before direct flights from Cape Town) led to renewed interest in seeing if it would be possible to send proteas and Cape greens by sea. SAPPEX, under leadership of

Barrie Gibson, together with Blue Star Shipping Line, and the Perishable Product Export Control Board (PPECB) cooperated to bring this off. It was of great help that Mr. Ginsberg who for years headed the cooling research at the Food and Fruit Technology Research Institute (FFTRI) offered his services. He was a consultant to PPECB at the time. After controlled condition tests, it was hoped to do a real sea trial for which Safmarine had already offered a container free of charge, no doubt with many containers of protea exports by sea in mind. It was estimated at the time that there would be a 50 per cent saving on freight if commercial quantities could be exported by sea. A sea consignment was simulated at Elsenburg. Flowers were put into a controlled environment for twenty-one days. It must have been very exciting to find that after twenty-one days most of the floral products were in a very good state. In particular the *Leucospermums* (pincushions) and the Cape greens looked very fresh. The proteas were not so good, due to leaf blackening, which was a problem in any case, also when sent by air. Magnifica also looked superb after a lengthy period of simulated sea freight.

In 1980, a number of proteas and Cape greens were sent by sea in a refrigerated container. The shipment left Cape Town on 30 June 1980 and docked in Rotterdam on 21 July. On first impression about 60 per cent of the shipment was satisfactory. Some items looked very good, while others were wet and were turning brown. There was some doubt whether the temperature had been properly controlled throughout the container, and it was thought that the way the container was packed might have something to do with some products looking better than others. Winston Odendaal and James Wood flew to Holland to meet the container that was shipped to Klaas van Zijverden. A meeting was held with all parties involved (S.A. Embassy and shipping line personnel) who agreed that unless the temperature of the entire container could be kept to 1.5°C for the duration of the trip, it would not be worthwhile to export containers by sea. Between Rotterdam and Aalsmeer, the container was not connected to a clip on refrigeration plant and the temperature rose to 10°C. At the time airfreight for one carton was R22.00, whereas by sea the cost was R8.50 plus clearing costs, which of course is a considerable saving. It would require some very good cooperation between suppliers, exporters and buyers to coordinate sea shipments as naturally a six meter container at a time is a different kettle of fish to a few pallets on an aircraft.

While there, Winston and James held a meeting at the South African Embassy in Den Haag, Holland where export regulations were discussed. Although those present realised that it was difficult to set out specific regulations, it had become necessary in order to protect the industry on the one hand and conversely to enable the Department's inspectors to carry out their duties efficiently. Also discussed was the brochure and the desirability to produce an informative and eye-catching poster for promotion of our products. James wrote in his report: 'I realise that we need a promotion with impact over a large field to sell our proteas abroad'.

On the research side, a lot of work was being done around the country and as is usual with research, each step forward resulted in further questions. For an industry that was so keen to make progress, the time-consuming and thorough steps required seemed to take too long. When Andries Claasen of Pretoria University reported that different protea species require different feeding regimes farmers were not exactly overjoyed, as they had hoped for something less complicated, but it did explain why some flower species did well under a certain program while others did not. Research was now being undertaken on soil requirements, leaf blackening and tissue culture.

Also in 1980, Indo Atlantic, the airfreight agents, withdrew their support of SAPPEX, believing that SAPPEX could now stand on its own legs, but pledged to continue to give financial support for production of the newsletter. Up to then they had paid a half-day Secretary, Reona Sivertsen to run SAPPEX affairs at their premises. To cover the cost of a secretary and the necessary equipment, the Executive Committee felt that they would have to raise the subscription from R15.00 to R25.00. They were well justified to make this increase — but for sponsorship the membership fee would probably have increased long before. First thoughts about a levy surfaced, but it was thought that it would not be feasible at this time. It was a worrying time for SAPPEX. Willem Verhoogt of Bergflora came to the rescue and said that he would make space for Mrs. Sivertsen in his office and that he would oversee her work, and he would not charge SAPPEX anything, except her salary.

The Editor of SAPPEX News, Mr. Harry Wood, resigned in March 1981. Reona Sivertsen resigned in September of the same year, after five years of service, and I took on the job of Secretary of the Association and Editor of the Newsletter, with Barrie Gibson at the helm as Chairman. I was thrown into the deep end with little guidance, a typewriter, membership index cards and a box full of files. One of my first jobs was to make a detailed submission on the Protea Industry to the Western Cape Agricultural Union, for which I relied on help from Walter Middelmann and Petrus Roux.

The industry had shown very good growth from 1975 and then suddenly there was an 11 per cent drop in 1981. Increased airfreight costs and the strong Rand meant that proteas were landing 30 per cent more expensive than in previous years. Perhaps as a result of this downturn, dissatisfaction arose between exporters and growers. Producers had the expectation that exporters had to sell all their production at the highest possible price; the exporters were accused of huge profits to the detriment of the producer. On the other side was the exporter who had the quality conscious buyer, the exchange rate, inflation and doubtful farming practices from producers to cope with. Producers were warned to get their act together, to harvest at the right time, to ensure 'clean' flowers and to maintain the cold chain, or face 'falling off the bus" Those who complained the hardest were the ones who didn't keep pace with new developments. It was an ongoing struggle that carried on unabated as long as veld harvesting was the norm.

John Rourke, curator of the Compton Herbarium at Kirstenbosch National Botanical Institute produced the beautiful book on 'The Proteas of Southern Africa' illustrated by Fay Anderson and Lura Ripley in 1980.

In 1981, South African Airways sported protea pictures on the menu card, done by local Kleinmond artist, Johan van Niekerk.

The Protea Research Unit at Riviersonderend released the Kouga variant of *P repens*, a deep red variety that flowered four to six weeks before the standard *P repens* of the area. At the same time a pincushion cultivar 'Vlam' was released. The mother plant from which this cultivar was developed was donated by Mrs. C.W. Green of Somerset West. Two further cultivars 'Helderfontein' (*Ls glabrum*) and 'Luteum' (*Ls reflexum*) were released for landscaping purposes, not for export production. Also in 1981, Dick Rust, entomologist at the Horticultural Research Institute at Stellenbosch devised a spray to be used for spraying into each flowers in order to kill beetles and mites before export. He found Dichlorvos to be very effective and it was found to be an excellent alternative for methyl bromide if used correctly. He counted over 3,000 insects per flower in random samples of twelve heads of *P repens*, while as many as 60,000 mites could complete their whole life cycle without leaving the flower, making it pointless to treat the flower from the outside. A spray program was designed by the Protea Research Unit against larger insects like miners, borers and beetles, but they warned against overdoing it, and suggested that farmers should rather concentrate on a good sanitation program.

From time to time, adverse press still appeared and one article accused Fynbos farmers of 'Rape of the Veld' and strongly condemned them for over-harvesting practices. Unfortunately, the **Sunday Times** in London picked up on this and published a damning report entitled 'Wild Flora Plucked to Extinction' and there was talk of a boycott of South-African indigenous flora. It seems this had resulted from a report submitted to the Flora Conservation Committee, which had been leaked to the press. It became quite a topic. Gert Brits wrote a long report to the Editor of SAPPEX News clearly outlining how little impact harvesting really has on the natural resource, and suggested that a survey be done in order to be able to refute such unsubstantiated claims. Until then the practices used had been developed by individuals, such as how to best cut/prune their wild-growing proteas, how much to leave for seed, how to pluck their Everlastings and how and when to burn for regeneration. There had been no coordination and no scientific investigation nor even collecting of data on all this. Therefore SAPPEX requested the authorities to assist the industry with a survey and research on veld harvesting. In 1983, the assistance of the Botanical Research Institute was sought, and eventually, after a long delay, an Industry Questionnaire was drawn up — known as the Eksteen survey. Unfortunately it came to naught, as the survey was never done

Dr. Gerard Jacobs (1983) delivered his now famous address 'The nursery — a bottleneck in the Protea Industry'. It took quite a few years before this problem was solved. In fact it took till 1999 and even then things went 'stop-start' for quite a few years thereafter. Anne Gray negotiated with S.A. Nature Foundation to match research funding for three years. In spite of this wonderful gesture, the members were rather backward at coming forward to put money into their industry. In the first year, members pledged R3,000 and by the following year it went down to R2,195 and probably even less the last year. However, in those days a little money went a long way and with the doubled money, SAPPEX sponsored research on Witches Broom and Flower Initiation studies as well as an investigation into vegetative propagation of proteas.

For some reason, growers did not want to get involved in SAPPEX, particularly not at the committee level. Accusations of being run by exporters, while true, was absolute the fault of the producers themselves for not coming forward to be heard. In order to encourage producers, they were invited to form a sub-committee with a representative to the main Executive Committee and a meeting was arranged for a frank discussion on costs and prices in order to foster a better understanding between exporter and producer.

Meanwhile, Proteas stayed in the limelight and in 1982, Brian Rycroft, in order to celebrate Kirstenbosch's 70th jubilee, conceived the idea of a huge flower show to be held in 1983. He was adamant that also cultivated proteas and Cape greens should be on show, and encouraged various towns in the Western Cape to hold their annual flower show at the event, which was to be held at the Good Hope Centre. Flora 83 was a huge success and is still talked about but sadly has only been repeated once since then.

Interest in protea-growing had of course long since extended to outside the borders of the Western Cape. There was very good success with plantings in the Krugersdorp district. Mount Sheba in the Transvaal did pioneering work up there to grow orchard type plantations on part of his 1,200 hectare private nature reserve. In 1982, the research unit therefore decided to have a farmer's day at Krugersdorp. About 75 people were expected but 200 arrived; such was the great interest in this 'new' horticultural crop for the Transvaal. After a series of lectures in the morning, with Gert Brits talking on propagation and Sharon von Broembsen on disease management, the protea farm of Gerrie Roos was visited in the afternoon. His cuttings house elicited much interest. The talk by Kleintjie Meyhardt on the advantages of cultivation in the Transvaal, away from the natural resource, would give rise to cultivar development and cuttings-based orchard establishment, which would be a clear advantage to growers there.

Already Gerhard Jacobs had warned the Western Cape growers that they should move away from seedling-based plantation in favour of cuttings-based plantation. In this regard, he warned the industry that it would stagnate if they did not look to the future and better quality and new varieties. He continued to warn the growers that production from the veld was inhibiting horticultural development.

In 1982, there was an attempt to form an export organisation under the name CAPE Associated Protea Exporters, based in Hermanus. The people who promoted this new organisation required a membership fee of R200 and promised to sell all products for a service fee of 5 per cent. In case of a claim, the producer would be fully responsible for the cost of the goods, airfreight and fees. There were a number of one-sided rules and regulations to benefit this organisation. It never got off the ground. Years later there was a similar attempt in Cape Town, which equally met with no success. Protea farmers aren't stupid!

Forced air cooling at Protea Heights. Kobus Steenkamp demonstrates. (Farmers Weekly Dec 1982)

Protea Heights, under management of Kobus Steenkamp was one of the first to install a cold chamber in the packshed. Flowers were packed in special cartons with holes. Once packed, the cartons were stacked under plastic sheeting with an extractor fan on one side to draw cold air through the holes in the cartons. Warm air was expelled from the cooling chamber, bringing the temperature down to between 1 and 3 degrees Celsius. After that the holes were sealed. The clear advantage was that instead of bringing small consignment to the airport's cool rooms, he could now take larger consignments less frequently, and by bringing the temperature down quickly, the flowers remained in a better condition for far longer. A further advantage was the delay in the dreaded leaf blackening that plagued so many proteas.

Then the dried flower exporters were confronted by the intention of the S.A. Dried Fruit Cooperative in Wellington (SAD), to enter the dried flower market. Some of their fruit supplying members owned property, which had considerable chunks of natural mountainside covered in indigenous flora. Some were already supplying dried flower exporters. They were, however, used to the cooperative system for the marketing of fruit and it for them it was of course advantageous to support the SAD on their doorstep. The dried flower exporters were invited to a meeting to discuss the SAD's proposed methods of operation who invited them on board to form a single marketing body. Naturally, the existing exporters had no intention of 'selling out' what they had built up over a number of years, particularly as they could only see difficulties ahead for a one-channel marketing operation. The 'godfather' of the dried flower exporters stated: *'They feel they would perhaps want to enter the Dried Flower business, but are unsure. So they try what Jan Haak tried: to scare existing exporters into entering some sort of combine with them, i.e. 'buy' them for their advantage. They would be willing to come to an arrangement re prices, i.e. they are in fact afraid'*, and regarding being sidelined by them if not willing to cooperate, Walter said, *'let them try!'*

And so the SAD became a role player on the flower scene. From 1983 when it made its first tentative moves, to full operation took some time, and after a number of years of operations they closed the doors on dried flower exports in about 2005. In the intervening years they had taken over a considerable market share and to the ire of the other operators instituted an 'agterskot' or back-pay system at the end of the season; being a cooperative, they were obliged to pay out surplus funds to their shareholders or members. What was not acceptable to the other dried flower operators was that SAD would cross-subsidise fruit profits also to the dried flower suppliers, thereby causing an artificial price differentiation to suppliers. This caused an upward spiral in supplier prices, which was difficult to recoup in a market that had a very low inflation and was not prepared to accept price increases. At the same time, SAD became a training ground for operators, with Johan Marais leaving them to run Bredaflor, a sister company of a German firm who wanted a direct foothold in South Africa in order to cut out the 'middleman'. Maans Wagener, first a dried flower supplier, became one of SAD's liaison officers, and later became a supplier liaison officer at Honingklip Dryflowers. Jimmy Wright, who took Johan Marais' place at SAD, later joined Pan African Corporation in Botrivier.

In 1983, Brian Rycroft retired after more than twenty-nine years as Director of the National Botanic Gardens.

Articles continued to be published about protea. In 1983, the *Farmers Weekly* published an article about Mr. Klasie Strauss under the heading 'Sing jy van blomme dan sing jy van geld' (if you sing of flowers, you sing of money). Of course, such reports encouraged still more people to climb on the bandwagon, which was not a healthy development. Klasie produced a number of crossings, which he registered internationally, amongst them being 'Ruth' and 'Sheila'. He also dried excess flowers for the dried flower trade, which at the time was about equal in value to the fresh protea export sales. Quantities of dried were of course much larger, considering that the price of dried flower was roughly one-third of that of fresh! The advantage in producing dried material was that it all went by sea. Reports were also published about production in the Eastern Cape. Production in that region was not new, as one of SAPPEX's founder members was Nokie v.d. Burgh of Kingflora near Thornhill outside Port Elizabeth, who specialised in *P repens* selections. Farmers started coming up with interesting information that they shared willingly with others. A small example was that of treating Serruria florida (Blushing Bride) seeds with milk to improve germination.

By 1983, SAPPEX listed nine dried flower exporters, eleven fresh flower exporters and six seed exporters. Renate Parsley had meanwhile taken over Ruth Middelmann's seed business. She published an informative booklet of descriptions of every species she sold.

The protea research unit at Riviersonderend released two new Leucospermum cultivars, namely L cv Scarlet Ribbon (*L glabrum x L tottum*) and L cv Sunrise (*L cordifolium x L patersonii*). These cultivars were made available to the trade as unrooted cuttings. Scarlet Ribbon was not very popular in South Africa until the Zimbabweans had huge success with it in Europe. To me it is still one of the prettiest pincushion cultivars. The unfortunate thing was that the trade wanted cultivars that flowered October to January, and most of the cultivars released flowered from August/September to mid-October. In order to shift the flowering time of these cultivars, debudding to delay flowering time would have to be done.

One of the problems with trying to export to the U.S.A. was the fact that all flowers had to be fumigated with methyl bromide, which unfortunately rendered the flowers unsalable as it completely destroyed the shelf life of the flowers. A local chemical firm recommended Formaldehyde spray and Para-formaldehyde granules. These treatments were apparently not successful. Therefore, the 'woolly' type proteas like *P neriifolia* and *P magnifica* were not good candidates for export to the U.S.A.

Another scandal erupted! An article appeared in *African Wildlife* that a rare protea, *P. holosericea* had nearly been picked to extinction deep in the mountains between Robertson and Worcester. First discovered by Scots botanist James Niven in about 1801 its location was 'lost' until it was rediscovered by Elsie Esterhuysen of the Bolus Herbarium in 1965. The farmer involved in this scandal appeared in court in Robertson and was convicted of harvesting protected wild flowers without permission of the landowner. African Wildlife Vol. 38 No 4 under the otherwise excellent article 'The Law of the Jungle' asked Walter Middelmann, as leader of the association and one of the 'culprits' (having bought the dried flowers) as mentioned in the story, for an explanation. He replied that the number of blooms plucked was exaggerated, but also suggested that to avoid scarce species from being utilised, the 'Endangered Flora' list should be updated. He as the SAPPEX representative on the Botanical Society's Flora Conservation Committee had already suggested this and his proposal had been accepted. Subsequently, a farmer in Barrydale successfully established 1,000 plants of Holosericae, which flowered four years after sowing. This incident led to further calls by a range of people for a proper survey and harvesting guidelines from the industry to be provided to botanists and conservation officers. A well-balanced article by George Davies appeared in 'Veld and Flora' under the heading 'Flowers from Fynbos — the need for a policy of resource management for the wildflower industry' did much to cool the situation and bring about reasonable conversation.

On a more positive note, SAPPEX published a list of all available research papers, seventy-seven in all. The old copies (published before the advent of computers) were subsequently captured and published in full. The Newsletter was upgraded to a Journal (issue 33-34 Dec 1984) with its own cover instead of the Association's letterhead.

Veld and Flora's December 1984 edition was devoted to proteas in preparation for the International Protea Association Conference and Protea Working Group Symposium that was held in 1985, with authors from as far afield as New Zealand and Hawaii supplying text. One of the authors was Marie Vogts who was, to the best of our knowledge, the first person to independently, many years prior to her appointment as researcher at FFTRI, undertake research on protea already in 1940, in spite of Prof. Compton telling her she was wasting her time. In 1958, after investigating fifty-two species representing eight genera, she had sufficient material and knowledge to be able to publish her book 'Proteas — know them and grow them'. In 1960, she made a public plea for research funding. Much to her delight the State appointed her in 1964 to assist a fledgling industry and she wrote: 'It is clear that had it not been for possible commercial value, research, essential also for the conservation of the original sources in the veld, would still have been sadly lacking'.

(Refer to Appendix G: Marie Vogts appointed as researcher)

(Refer to Appendix H: Research into the South African Proteaceae, Marie Vogts)

In 1985, I resigned as Secretary of SAPPEX and became a fully fledged member of the Executive and was officially made Editor of SAPPEX News after already having done this job since 1981! After a short-lived appointment of a new Secretary in Somerset West, Paloma Werner took over as Secretary/Bookkeeper. Pretty soon a number of male members of the association were talking about learning Spanish! Paloma who had been a Secretary at Honingklip and missed her job since staying at home raising her babies, ably coped with this post as a part-time job while raising her youngsters.

A Fynbos Biome Research Meeting was held in Stellenbosch in July 1985 where the first survey on the industry was presented by Melanie Simpson of the Department of Nature Conservation under the title 'Survey of Indigenous Flora utilized by the Wildflower Industry'. To this day, it is an excellent reference work; particularly useful were the common names given to plants in various regions. According to her report, more than 300 indigenous species of the Cape Floral Kingdom are utilised by the flower trade in fresh and/or dried form.

SAPPEX celebrated its twenty-first birthday at the farm of Petrus Roux, complete with a birthday cake cut by Anne Gray.

Because of the problems experienced with sufficient freight space from the Cape in the summer months, SAA started to operate a refrigerated road service from Cape Town to Johannesburg airport.

International Developments

International Protea Association

The International Protea Association met for the first time in Kallista, Victoria, Australia from 4 to 8 October 1981. Main speakers were Dr. Phil Parvin of Hawaii, Johannes van Staden (University of Natal) and Gerhard Jacobs (University of Stellenbosch) of South Africa, Frank Greenhalgh, Burnley Plant Research Institute, Vic, Australia, Mr. Edward Muzzy of California and Walter Middelmann of South Africa who spoke on the development and structure of the South African Protea Industry. This talk, because it really gives the 'state of the protea nations' as at 1981, is reproduced in full.

(Refer to Appendix I: Development and Structure of the South African Protea Industry)

At this meeting there were 116 registrants, representing every major protea-producing region in the world, and they adopted an interim constitution for the new association. Rather than recognise political boundaries, the conference decided to recognise as many different production areas as the delegates decided upon. The delegates grouped themselves into geographical production areas and each group selected a representative to sit on the interim Council. The Council would elect the Officers (President, Vice-President and Secretary/-Treasurer) and an Executive consisting of the Officers plus two other members of the Council. The officers and executive would be responsible for managing the business of the association between conferences and would produce an annual report and financial statement.

The formation of the association was even reported on in South Africa together with a picture of some of the important role players who had previously visited South Africa and were seen talking to N.C. Krone of Twee Jongegezellen at the Tulbagh flower show.

*International visitors after first gathering
of Protea people in Australia
fltr: N.C. Krone, South Africa; Peter Mathews, Australia;
Barbara and Ray Schatz, California and Phil Parvin, Hawaii
(Landbouweekblad 16 Oct. 1981)*

Joy Amos of New Zealand, a local extension officer in Auckland was elected vice president of the newly formed IPA. She wrote to Ruth and Walter: *'May I first congratulate you on becoming Founder Members of the International Protea Association. It seems most appropriate in every way, as so many people in New Zealand, and no doubt in many countries first started growing proteas with seeds from your nursery. How well they have grown! Peter (Mathews) did a good job of editing the proceedings of the conference — so much was packed into that small volume. Phil Parvin keeps in touch and the response to plans to hold the next IPA conference in conjunction with the Society of American Florists conference appear to be well received. It's a great opportunity for growers and exporters in this country and the conference should be a further step towards coordinating supplies worldwide — an exciting concept though a mighty difficult one to achieve!'*

In 1983, the conference was held in Hawaii (Technical Seminar) in Seattle (Marketing Seminar) and this was followed by a Protea Field Tour from Los Angeles to San Diego. Barrie Gibson was sponsored by SAPPEX to attend this conference as he would be the next president (the next host country's leader always assumed the title of president) He presented a short film, borrowed from Kirstenbosch, on South Africa as an introduction and to serve as enticement to join the South African IPA Conference in 1985. At the same time Gert Brits was sponsored to go to Washington to sort out problems regarding insect and disease inspections that prohibited South-African proteas from entering the U.S.A. and from there he and his wife, Cecilia, joined the conference. It was their first trip to the U.S.A. and his first contact with the IPA where he was soon recognised as a leader in the research field. The Seattle seminar included sessions with major U.S. wholesalers in the morning and protea design demonstrations in the afternoon after a joint lunch with members of the Professional Floral Commentators International. The trade fair area also drew a lot of interest with proteas in full force in displays by New Zealand, South Africa, and Australia, while the Hawaii display featured dendrobiums, antheriums, and exotic foliages in addition to proteas.

The IPA ran a brainstorming session on Promotions, including providing proteas for use in design schools, production of a catalogue and wall chart. They also discussed fund-raising for research and marketing. An international research committee consisting of participants from the various attending countries was elected. They were to concentrate on plant breeding, diseases and pests, post harvest, nutrition, pruning, and propagation. For forthcoming conferences, area representatives were asked to profile not only the number of growers and extent of plantings, but also to give a report on research undertaken.

The first IPA Newsletter came out in October 1983, with Mrs. Joy Amos as Editor and the first research papers were published as early as August 1984, kicking off with 'Know Your Proteoid Roots' by Gert Brits. A protea poster was developed and was for sale internationally. It featured both the South African and Australian genera. It might be said that SAPPEX

News was a good source of information for the IPA journal and vice versa, while all other Protea Associations also benefited from exchanges that were freely available between all these publications.

The International Protea Working Group (IPWG)

Phil Parvin and Dr. JT (Kleintjie) Meynhardt were the main movers for the formation of the Protea Working Group. In August 1983, Phil briefed Kleintjie on progress since their discussion in April. On record is a letter from John Seeley, President of the American Society for Horticultural Science in Virginia, U.S.A., responding to a request by Phil Parvin in which he confirms that he will write a letter to support the formation of a Protea Working Group sponsored by the Section for Ornamental Plants of the ISHS and he enclosed the letter addressed to Dr. Van der Borg, Secretary General of the ISHS. I quote from this letter:

At a meeting of our Section Members at the time of the Symposium of on Production Planning in Glasshouse Floriculture in Denmark next month, we will establish a Working Group for Proteas and then will select a chairman to serve until the time of the IH Congress in Davis in 1986.

To this end, scientists were invited, at short notice, to attend a preliminary meeting to be held in March 1984 in Cape Town. Scientists from around the world responded. A SAPPEX subcommittee, of which Anne Gray, Barrie Gibson, and Walter Middelmann were the main movers, secured funding for visiting scientists and their accommodation to supplement partial funding by the Horticultural Research Institute (HRI) for this very special meeting.

The agenda covered reports and discussions on:

Organisational matters: objectives, communication, and election of office bearers

Status of Protea Research and Cooperative Research

Field Trips to Protea Heights and Jonkershoek, Tygerhoek, and Mountain Range Flora

During the group's overnight stay at the Birkenhead hotel in Hermanus, SAPPEX members were invited to meet the scientists. Dr. Phil Parvin (Hawaii), Dr. David Dennis (NZ), Mr. David Nichols, and Dr. Byron Lamont (Australia) each gave a short talk. It is worthwhile to record the scientists who attended this historic event:

Dr. TJ Meynhardt (HRI, South Africa)
Dr. GJ Brits (HRI, South Africa)
Mr. JH Coetzee (HRI, South Africa)
Prof. PS Knox-Davies (US, South Africa)
Dr. G Jacobs (US, South Africa)
Prof. G H de Swardt (RAU, South Africa)
Prof. J van Staden (UN, South Africa)
Prof. J Eloff (Director, NBI Kirstenbosch)
Dr. PE Parvin (Hawaii)
Dr. D Dennis (New Zealand)
Dr. N Nichols (Australia)
Dr. B Lamont (Australia)

The first newsletter of the IPWG, aptly named 'Protea News', appeared. The editor was Adele Nortjé of the Horticultural Research Institute, South Africa, with Phil Parvin of Hawaii as first Chairman. In the first Editorial he wrote:

A historic event took place in the Cape Province, March 26-29, 1984. Culminating months of activity, the Protea Working Group, Section Ornamental Plants, International Society for Horticultural Science was duly organised. Tracing its origins back to conversations between the Director of the Horticultural Research Institute and the President of the International Protea Association in 1982, general agreement was reached on the value of getting research scientists with interest in proteas together to discuss their progress and problems. In 1983, the Associate Director of HRI recommended to a visiting researcher from America that the recognition of a Protea Working Group within the ISHS had advantages in facilitating communications between international scientists. A petition was drafted, presented and approved at the meeting of the Section Ornamental Plants, ISHS, in Nyborg, Denmark, August 17, 1983. The minutes of this meeting record that the formation of a Protea Working Group was approved.

The organisational meeting was held in Devon Valley, Cape, RSA, March 25-29, 1984, hosted by HRI. Representatives from Australia, New Zealand, United States, as well as three Universities, the National Botanical Gardens and the Horticultural Research Institute made up the organizational group. The name and objectives of the group were ratified, the first slate of officers were elected, the dates and general format of the first International Protea Research Symposium were determined, and a symposium committee approved. The IPWG promises to be a dynamic force in the development of information upon which a new international cut flower/cut foliage industry is based.

Welcome Aboard!

In newsletter 2, information was published on the International Cultivar Registration system and checklist for Proteas, which was the responsibility of Gert Brits of South Africa, while Australian genera registration was the responsibility of the Australian Cultivar Registration Authority.

The 1985 IPA conference organised by SAPPEX was very important as the ISHS's International Protea Working Group (IPWG) and the IPA worked together for the first time. IPA would continue to host the IPWG at every second conference. Dr. Kleintjie Meynhardt, then director of the Horticultural Research Institute pledged his Department's support for the meeting. The event was attended by Mr. Van der Borg, Secretary General of the ISHS.

Kleintjie Meynhardt

The IPA Conference and IPWG Symposium were held in the Transvaal and in the Cape, with a full conference program as well as visits to protea farmers, bulb growers, and to Multiflora in the Transvaal. A host of activities took place in and around the Cape, where Walter was responsible for a unique exhibit at Cape Town's national library in 'The Gardens', under the name 'Proteas in Picture and Print' showing newspaper cuttings, old manuscripts, books, and botanical art. Walter put up quite a few books from his own collection, the most noteworthy one being 'Clusius 1605' showing the first illustration of a protea. A visit to Kirstenbosch was of course included. This event got lots of publicity in the local press as well as in agricultural magazines, particularly since the international scientists were talking about new breakthroughs like tissue culture for proteas. It came as a big surprise to many South-African growers to learn that proteas were grown in faraway places like Israel and Hawaii and there were protests of 'selling our birthright' and worse.

In spite of adverse press overseas of the political situation in South Africa, 200 or so delegates, many of them from Israel, U.S.A. and Australia came to South Africa and found that their experiences were totally different from what they had anticipated, with the much publicised political unrest not apparent anywhere. No incidents occurred of course and the visitors left with a very different view of the country.

At the Cape Town conference, the IPA decided to allocate A$10,000 towards a joint SAPPEX/IPA project to produce a new protea marketing brochure.

New Zealand

Dr. David Dennis at Levin's Horticultural Research Centre was responsible for supplying an advisory service to Leucadendron growers and providing answers to their questions. 'There's been a mad bonanza with these plants, next year about half a million blooms will be ready for sale' said Dr. Dennis. He predicted that by 1986 New Zealand could be producing four to five million Safari Sunsets bringing in approximately $1 million NZ.

In 1983, Lewis J. Matthews of Canterbury in New Zealand published his 'South African Proteaceae in New Zealand' with paintings by Zoe Carter. It was probably the first publication on the cultivation of proteas in New Zealand, although nurserymen in New Zealand had pioneered vegetative propagation of proteaceae in large numbers since the 1940s, and a remarkable number of selections had been named and marketed.

Jack Harré submitted a report for the IPA on Proteas in Commerce in 1984 and wrote in the closing paragraph: '*As a commercial flower crop Proteas are today where roses were fifty or more years ago and orchids, antheriums and strelitzia were fifteen years ago. These flowers and many other have all gone thru their stages of development in plant breeding, clonal selection, post harvest research and the matching of the crop to the market. Protea have yet to begin such a program in earnest'.*

In 1985, the Ministry of Agriculture and Fisheries produced a very detailed leaflet on proteaceae flower and foliage production: varieties, cultivation, and harvesting. The protea world was definitely beginning to expand on all fronts.

Japan

Small experiments were on the go in Tokyo, with Mr. Shigeyuki reporting that he had a mere 200 square meters of proteas in bags under cover. The *P neriifolia* grew very tall under those conditions, similar to the nutrition experiments that took place much later in France. In 1982, a Japanese flower buyer visited South Africa and made contact with the Middelmanns. I remember vividly that it was a miserable and cold day. They arrived here in short-sleeved shirts. Walter, who used to have a short nap every afternoon, could not find them after his nap. Robert and I were busy elsewhere and hadn't seen either of them. Eventually after searching high and low, it turned out that they had taken themselves to bed because they felt so cold. They had fallen asleep in the process! Afterwards, they sent us their fresh flower brochure showing many South-African species with the cover page featuring a large *P. compacta* bush, which they had photographed on the farm.

In 1984, Walter supplied text on dried flowers from Africa to Mr. Hirao in Tokyo for an article in *Garden Life*, published in its 1985 issue. In it he traced the origins of the South-African 'natural dried decorative floral materials' comprising not just proteaceae, but also branches, grasses, vines, and cones, all of natural origin. The article featured a number of colour photographs, mainly taken at Honingklip Farm.

Australia

In 1980, the Department of Agriculture in Victoria, Australia presented a mini-symposium and published notes on 'Proteas for Profit'. At the time there were about 100 ha under protea in Victoria State. This would be too much for the local market and it was recommended that growers should market the blooms in Japan, Canada and U.S.A. A marketing cooperative, Protea Australia, had been established with about 80 members, who after a short time in business, had run up a considerable operating loss.

In 1981, the Federal Government spent R76,000 (Cape Argus report) on a study into the export potential of the Australian flower market. The report states that South African proteas were the most popular foreign flower being grown in Australia. It is interesting how difficult it can be to make correct interpretations from reports and articles. In an article in the *Australian Financial Review*, it was stated that 'Even though South Africa is the home of the protea, none are grown there for export. Hawaii is the biggest world producer of the plant at this stage'. Yes, at the time not many South-African producers were growing proteas in plantations, but rather were harvesting from the veld, but a reader without proper background knowledge would imagine that South Africa was not even considering export at that stage, which certainly was not true.

In 1982, the Australian Department of Fisheries and Wildlife in report no. 53 entitled 'The Western Australian Wildflower Industry 1980-1981' wrote: 'It is stressed that the conservation problems posed by harvesting for the wildflower industry probably are negligible compared with those arising from total or partial destruction of the native flora by clearing for agriculture, public utilities, mining, prescribed burning and urban development'. The parallels between South African and Australia are obvious.

Scams and misrepresentation of facts are nothing new to the world. A circular letter from the Tax Investors Club in NSW, promised a 100 per cent tax deductable investment, with 100 per cent p.a. returns after five years, with returns possibly reaching 400 per cent. There were wild statements like *'Even if $1 billion was invested for growing flowers, Australia could still not meet all the demand. The market is there for the picking. Among the most valued flowers on the massive overseas markets are the Waratah and the Protea. Both can be grown in Australia. In fact, Australia is almost the only country in the world that can export them because imports from South Africa are banned by many countries'.* The offer was for 120 units over 200 acres and asked potential participants to 'Immediately priority post us your initial $2,500 and power of attorney'. The whole matter was referred by suspicious growers to the Minister for Business and Consumer Affairs who referred it to the Trade

Practices Commission. Unfortunately the rest of the documentation is not available to me, so how this saga ended is unknown, but I am willing to bet that the Tax Investors Club was not successful.

A more balanced article, 'Horticulture in Australasia' written by Peter Mathews appeared in *Greenworld*, which featured proteas on the cover. Peter's history and entry into the world of proteas is a very interesting one and appeared as an article in *Prime Time*, April 1983. As a teenager Peter was an accomplished musician and played violin with the Adelaide Symphony Orchestra, but instead of taking up music full time, he got into youth work and for ten years was Director of Youth Work in South Australia, including a spell as Federal Director. When their four children were still really youngsters he was asked to go to Africa to work in what was then Northern Rhodesia. He became a founder and director of the Mindolo Ecumenical Foundation. The 'revolution' was brewing with the black people wanting their independence from colonial power. He said 'it was a bit of a wild scene at times'. After six years he returned to Australia and became director of Australian Frontier, an organisation formed to bring people together to discuss the major issues facing Australia.

Peter Mathews (Australian Horticulture, Nov 1981)

It was while working in New Guinea for *Australian Frontier* that a chance meeting with a woman started him on the road to protea growing. She said she had a property near Melbourne and was looking for a plant that was low maintenance. Peter offered his assistance and in his hunt came across the protea. He watched progress on her property and said: 'my giddy aunt, they grow like weeds, I think I'll plant up my own'. He was then fifty-four years old and five years away from retirement.

He and his wife Rita bought a seven-acre property at Sylvan in the Dandenongs. At first they sold a few plants from the tiny nursery they established, then a few hundred . . . then a few thousand. He told his family that at the rate business was going he would soon cover an acre of ground. His sons and daughter thought it a huge joke, but a few years later the joke was on the family. They saw that business was booming and David, the town planner, Andrew the teacher, and John the mathematician all left their professions and bought blocks of land. Eventually, the whole lot became one company, Proteaflora. He felt the future was in hybridisation and of the various ones developed by him, the best-known is the ever popular Protea cv Pink Ice. Then followed training courses for prospective growers, lectures, and of course, the world conference that he organised, which led to the formation of the International Protea Association. Today, to my mind, Proteaflora is still the benchmark for protea nurseries. Arnelia Nursery (Hettasch) in Hopefield, South Africa, has become a close second! (I do apologise if there are others as good — they do not get a mention because I have not seen them.)

In 1984, the Queensland Protea Association published its first newsletter. The first president/treasurer of the newly formed association was Christopher Boast, with Judy Moffatt as Secretary. Within the year, the president resigned due to business commitments and his place was taken by Brian Richards. The Protea Association of New South Wales, published 'Grading and Handling Standards' in Newsletter No. 3, volume 3 of June 1984. In Western Australia, another association had sprung to life, 'The Commercial Protea Growers of Western Australia'. Elizabeth Wood, a relative newcomer to Australia produced their newsletter. James and Elizabeth had moved to Busselton Australia from Stanford in the Western Cape in January 1984 and within a few years made a name for themselves for quality flowers and quality service and went on to win a number of local awards. Barely a year after settling in Busselton, James and Elizabeth Wood had established 15,000 plants in their shade house, to be added to the thousands already growing outdoors. They encouraged others in their area to strive to breed and develop new cultivars uniquely Western Australian and to ensure that they propagated only disease-free material.

Peter Mathews, near Melbourne, continued to run seminars and even arranged a four-day Protea Festival at Ferny Creek Horticulture Society Centre, with proceeds of the first day's takings donated to the Freedom from Hunger Campaign. A few months later, a conference was held at Protea Panorama, with international speakers, and held under the auspices of the IPA. He also organised a National Protea Growers

Conference on behalf of the IPA Australian Chapter. I do not believe that any of the other countries ever did this. Whenever there was an IPA issue, the local Protea Association would take it up for discussion with its members, or they would notify their members if there was an international personality visiting so that local members could meet them.

After different associations had been established in the various states, a National Conference and general meeting was held in May 1985 in Sydney with participants from Queensland, NSW, and Victoria. The Australian Protea Growers Association was launched with Keith Bottomley of Kurrajong Heights as Chairman. Because of international trade beginning to turn against South Africa, the conference agreed to identify protea grown in the country as 'Australian Protea'.

Hawaii

The protea industry in Hawaii, which was established in 1975, made rapid strides with lots of state support. Most of the Hawaii production was sold to the mainland, apart from boxes purchased by the tourists. It is therefore not surprising that they ran a good advertising campaign. A lovely exotic looking poster was produced entitled Hawaiian Protea — intimating that this was a Hawaii exotic! The local Aloha airline featured proteas on the front page of their April 1982 in-flight magazine, complimented by an illustrated article on the proteas of Maui. Further articles were published in other Hawaii-based magazines. The firm Protea Gardens, Maui provided a leaflet with care instructions that accompanied their proteas. Suggestions included re-cutting on arrival and pulsing with preservatives or sugar and that florists store the bulk of their flowers for later use in a cool room. Innovative marketing led to sales of dried protea blooms, carefully preserved in clear plastic tubes, fifty-four tubes to a case, and each tube containing a small history of the protea, plus a four-colour photograph.

Exotic Blooms from Hawaii. Protea poster.

A South African visitor to Hawaii was concerned that the lucrative trade in South Africa's national flower was on the point of being hijacked by a country at the very opposite end of the earth and this resulted in an interview with the *Sunday Times*. While critical of production outside South Africa, he was impressed by the imaginative names given to the Hawaii proteas, which he reckoned should be emulated by South Africa. The journalist contacted other SAPPEX members for their opinion. Chairman Barrie Gibson responded that we were way behind in marketing ideas. Mr. Walter Middelmann, who had exported the seeds in the early 1960s for trials on suitable growing sites and from which the Hawaii industry was originally developed, stated that American claims of 'millions' were as extravagant as their advertising, while an exporter stated that the European customers would not take kindly to 'fancy names' after having got used to using mostly botanical names. What many people forget regarding such 'competition' is that the Northern and Southern Hemisphere production is six months apart and complement each other, rather than being in competition with each other, particularly in respect of *Leucospermums*.

Phil Parvin reported on the current status and potential of the Protea Industry at the Fertiliser and Ornamentals Workshop in Hawaii. According to him already by 1983, *Leucospermums* were the second biggest seller of floriculture products after *Antheriums*. In Hawaii charity funds were raised via proteas. In Maui, a local artist, Cynthia Conrad, made an original watercolour available for posters, to be sold in aid of 'Maui United Way'.

Such was the interest in South-African proteaceae, that the SF&N's Spring Buyers Guide of 1984 published a special section on South Africa. The reader was told '*As a resource for new plant material, South Africa has few equals. As a potential threat to the U.S. floral industry, it has a long way to go*'. But, of course there were opposing views too. Bradish Johnson of Protea Gardens, Maui, wrote to Washington that: '*South Africa was dumping flowers on the American market at prices far below what is possible for the American farmer to survive on*' and worse '*their shipments are treated with what seems to be an apparently dangerous chemical*' and '*there is considerable concern as to its long-term effects on the thousands of American florists who will be exposed and handling the South African Protea material*'. A complete exaggeration of course!

In 1985, an analysis of the protea industry and action plan for the future was done in collaboration with growers and the College of Tropical Agriculture and Human Resources. A number of bottlenecks were identified and solutions suggested. How many were acted on is not clear, but State support was certainly forthcoming for the Hawaiian protea industry. Under direction of Phil Parvin, the Maui research station produced some wonderful *Leucospermum* cultivars. On my first visit to Maui research station, I was amazed at the large number of pincushion varieties they were breeding, some with the most exquisite colours and shapes. They certainly were way ahead of anyone else.

England

On the occasion of the wedding of Prince Charles and Lady Diana, 120 'blushing brides' (*Serruria florida*) were flown to England by courtesy of SAA and were arranged by Joan Pare who had flown to London for the event. According to the Cape Argus, the South-African blushing brides stole the show. In 1947, Joan had been chosen by Interflora to present a posy to the then Princess Elizabeth when the Royal Family visited Cape Town in 1947 and since then and until late in life she continued to send proteas for royal occasions.

In 1982, the South African stand at Chelsea again won the coveted Wilkinson Sword and Gold medal. Proteas were also featured at the World Travel Market, opened by Princess Alexandra. More than 1,400 exhibitors from 90 countries were present and the big South-African stand was flanked by individual South-African firms. This exhibition created huge interest and certainly helped to entice more tourists to South Africa.

Proteas kept winning awards. In 1983 South Africa again walked away with awards, not only in Chelsea, but also in Edinburgh where a smaller version of the Chelsea exhibit won the top prize with floral material supplied by Honingklip Dryflowers and Klasie Strauss.

Israel

On record is a letter of 1981 from Dr. Michael Avishai, director of the Hebrew University of Jerusalem, who, with a friend from their Faculty of Agriculture, submitted a research proposal concerning the introduction to Israel and the trade of indigenous South-African ornamentals. Years later, South Africa made an agreement with Israel to provide cuttings material of new cultivars, much to the astonishment of SAPPEX who had not been part of the discussion. I was later told the reason in confidence (and so it shall remain).

Both the *Farmers Weekly* and *Landbouweekblad* in South Africa featured articles on proteas in the Holy Land during 1983. With their alkaline soils, *P. obtusifolia* was the favoured protea, although *Leucadendrons*, particularly Safari Sunset, were propagated under drip irrigation.

The developments in Israel were receiving much interest, and in 1985, Dr. Dawie Ferreira of the Agricultural Research Institute in Roodeplaat spent a year's study leave at the Volcani Institute at Bet Dagan, where he made a study of various biotechnology techniques.

U.S.A.

Rainbow Protea continued to produce very nice coloured brochures. They probably were not the only ones. Articles continued to be published, examples being one in *Florist Review* of 1983 under the sub-title 'These exotic, long-lasting flowers are fast gaining popularity in the fresh flower retail market' and another 'Proteaceae: from antiquity to American marketability'.

An amusing story by Ray Collett in Pacific Horticulture about protea growing needs to be retold:

Our first African proteas, which were chiefly members of the genera Leucadendron, Leucospermum and Protea, resulted from efforts to grow them from seed in a glasshouse. Most seedlings died, despite efforts to keep them free of fungus diseases. A few survived. These first survivors were never better than nuisances. Accordingly, one rainy autumn day, they were consigned to the cow pasture. They were miserable things, and no one wanted to give them any more attention. Out in the pasture, however, without attention, without warmth and without fungicides they began to show signs of health. Within a short time they recovered from their straggling lankiness and began to overtake the pasture weeds that grew around them. The cows sniffed at them with disdain, and soon we were watching the first stages of the development of the prodigious inflorescences for which proteas are famous.

According to the Arboretum Associates Bulletin of June 1985, some protea relatives contain cyanide-producing compounds. Wendy Swanson from University of California, Davies, investigated a number of plants at the Arboretum and found a number of cyanide-producing plants; these included *Grevillea, Hakea, Lambertia, Leucadendrons,* and *Telopea* — no wonder cows are not interested in eating proteas! (Quantities of cyanide are probably minute otherwise her report would have received extensive attention!)

Zimbabwe

Unfortunately, I did not get much information on Zimbabwe, although there is a cutting from the Manica Post in 1983 in which John Meikle, President of the Zimbabwe Protea Association, stated that Zimbabwe had favourable growing conditions for this crop. In the article it was mentioned that at least twenty varieties were being grown in the Vumba, Penhalonga, Inyanga, and other farming areas in Manicaland, and had been shipped to Holland in two trial consignments. Buyers there made favourable comment on their quality, and regular small consignments were sent from 1984 onward. An advantage for Zimbabwe was that the proteas flower there in summer, the European winter, and they were thus well placed for the window of demand in Europe. Because of relatively low establishment costs, it was thought that it would be an ideal crop for communal farmers. Zimbabwe also had the benefit of the Lome Convention with lower import duties and lower airfreight rates into Europe.

El Salvador

The first enquiry from that country came in 1985 from Mr. Veltman. Mr. Veltman, who also had an operation in Miami, soon established some hectares of protea plants for importing from there to his operation in the U.S.A.

Industry developments 1986 - 1989
South Africa

The industry exports reached 2000 + tons for the first time. So the industry was growing nicely.

Ruth Middelmann's seed business had been in the hands of Renate Parsley for a few years already, selling under the name 'Parsley's Cape Seeds'. She expanded the seed business to also include other indigenous Fynbos seeds and made a name for herself as a pelargonium specialist. She sold not only to growers overseas, but also to botanists, gardeners, and academic institutions such as universities and botanic gardens, often for experimental and research purposes. Another role player was Mr. Parkes of 'Feathers' Nursery who exported proteaceae seed worldwide, as did the Van Heerdens of Akkerdraai Nursery in Stellenbosch.

An article on nursery seed suppliers appeared in *South African Garden & Home*, August 1986. In this it was mentioned that surplus seed from the National Botanic Gardens had been available to members of the Botanical Society of South Africa since 1921. Although, at first, the selection was limited, the variety gradually increased and by 1986 the seed lists offered 673 species.

In the Argus of 1986, it was reported that the Cape Department of Nature and Environmental Conservation at Jonkershoek was in the process of computerising the records of threatened, rare, and endangered species. According to Ruida Pool, a botanist working on the project, almost a third of the proteas were threatened and almost a third of that occurred on private land. And so the pressure to be more aware of the environment they managed increased on farmers, and SAPPEX did its best to convince farmers of their responsibilities in this regard.

Protea species were voted to be the tree of the year in 1986. According to the booklet produced on this occasion, twenty-six of South-African protea species are classified as trees. One of them, *P roupelliae*, was depicted on the front page of the booklet. The largest concentration of proteas is found at the southern point of Africa extending approximately from Nieuwoudtville in a curve along the coast to Grahamstown. Sixty-nine protea varieties occur in this area, commonly known as the Cape Floral Kingdom. Thirteen other varieties occur in the rest of South Africa and further north in Africa another thirty-five varieties, making a total of 117 varieties in Africa.

Sharon von Broembsen of the Plant Protection Research Institute was the bearer of bad news to the industry. From her we all learnt about diseases on protea, as if the bugs didn't give enough trouble! It was a subject that had not really been researched on proteas and within a short time, growers got used to hearing about diseases such as *Phytophthora, Botryosphaeria canker,* and *Drechslera*. Most of her trials were done at Protea Heights in Stellenbosch and Mountain Range Flora in Kleinmond. She went off to America in 1986 to deliver a paper on 'patterns of invasion of natural ecosystems by plant pathogens'. She also visited plant pathologists in Hawaii and California to study *Phytophthora* root rot on proteas, forest trees, and other crops. It was fortunate indeed for the industry to have her on the team.

Cobus Coetzee appointed as Entomologist 1986. (Agricultural News 11 July 1986)

Meanwhile, Cobus Coetzee had joined what was now called the Vegetable and Ornamental Plant Research Institute (VOPRI) and took over as entomologist where Dick Rust left off. He taught the industry about borers, snout beetles, mites, and many other bugs that invade the flowers. According to him the best prevention was proper sanitation, which entailed removal and destruction of infested material and manual removal of larvae. Farmers needed to pay regular attention to their plants and pick at the correct stage before the invasion of nectar-seeking insects. He divided insects into three main groups: leaf-eaters, borers, and visitors.

Then pruning came under the loupe — this relatively new technique was expected to boost production and to increase the bearing life of plants. Again, this could only be done in proper plantations so the farmers who were serious about protea farming started planting and taking note of what the researchers said. At Protea Heights, Dr. Jacobs and Kobus Steenkamp and a host of students did a number of pruning trials. The results looked very promising and were instituted with excellent results. Another step forward had been taken towards professionalism.

Mrs. Bain of Wellington developed a cottage industry on her 30 ha smallholding, growing and selling proteas and developing dried flower posy gift boxes amongst others. She also did pottery, patchwork, and wool spinning. Her main outlets were the fresh and craft markets. Undoubtedly, there were others like her, who in a small way contributed to the industry as a whole.

Anne Gray took on Chairmanship of SAPPEX with Petrus Roux as Vice Chairman. Anne obtained a B.Sc., at Rhodes University, with Botany and Entomology as majors. She was actively involved in the dried flower business, which she ran from the farm Knorhoek in Sir Lowry's Pass. She experimented with colouring the dried material, which at first was not very popular on the overseas market but later was quite acceptable and now it is something done quite freely throughout the trade. She opened a Dried Flower shop in Somerset West. Other's like Valda Wegener of Newlands followed suit.

Petrus Roux was a progressive protea farmer in Kaaimansgat above Villiersdorp. A producer sub-committee was formed under his chairmanship. Petrus had a booming voice and when he spoke people listened. For years he quietly supplied proteas for Agricultural Congresses held in the Cape. Without fanfare, he made a donation to the ARC to establish a protea cuttings propagation facility at Elsenburg. Being one of the most liked characters in the industry, he started collecting old farm implements and now has a huge collection of old tractors and ploughs, which are housed on his property in Villiersdorp. Retired now, he still frequently writes letters to **Landbouweekblad**.

The Transvaal was becoming an important growing area and scientists went up to talk to them about diseases and insect control in the summer rainfall region. Anne encouraged the Transvaal members to form a subcommittee and there was a certain amount of interest, but for quite a while nothing more happened.

In order to encourage farmers to grow proteas as a crop, scientists and officials from the Department of Agriculture concentrated their talks on overseas developments and how our proteas could be 'plucked away by foreign competitors'. This of course drew adverse reaction from people outside the industry, who objected to our floral wealth to be traded worldwide. The South-African producers were by and large not yet ready to buy cultivar material as it was still too easy to harvest from nature, but as transport and labour became more expensive, it gave rise to acceptance to cultivate for ease of harvesting.

Although the industry did no direct marketing for its members, there was always some show or another where the government asked for proteas and growers gave willingly, or sold flowers to the government department concerned. In 1985, thirty-three companies from different industries were represented at South African Weeks held in Taiwan over the Christmas period. It was the largest promotion of its kind for South Africa because the Far Eastern Department Store not only exhibited the goods, but South African goods were also sold to the public. Proteas were very much in evidence in the general decorations.

A completely different kind of promotion was undertaken by Joan Pare, who was invited to prepare arrangements of Proteas for the wedding of Prince Andrew and Sarah Ferguson and to supply flowers for the wedding in Westminister Abbey. She also presented a basket of flowers to the Prime Minister, Margaret Thatcher, and Joan was delighted with the letters of thanks that she received.

And then boycotts. A report in the Cape Argus of August 1986, based on a report in the *Wall Street Journal* of New York, stated that 'in the US these days South African sport, South African art, even South African wine are tainted by politics' The report continued that South Africa was ranked 8th out of sixty-eight countries shipping flowers to the U.S.A. Perhaps just happenstance, or perhaps because of political lobbying against apartheid, imports from South Africa were carefully scrutinised. In November, the U.S.A. Department of Agriculture in Washington informed Pretoria that as of December 1, 1986, cut flowers of all species of Proteaceae from South Africa will be prohibited entry into the U.S.A. Similar action was taken against these flowers from Swaziland. My personal opinion is that this ban was partly due to the continuing high rates of interceptions of agricultural pests requiring treatment, destruction, or re-exportation. Meanwhile, the Dutch continued to export proteas to the U.S.A. and as one New York florist said 'You have to be deaf and blind not to know they are from South Africa'.

In the Netherlands, there were also calls for sanctions with wild and judgemental statements appearing in the press. Fortunately there were also slightly more moderate reports, particularly with regard to interviews with two of the largest importers of South-African proteas who emphasised work opportunities made possible by the flower trade. However, the anti-apartheid lobby was very strong and the upshot was that various Dutch stores like Albert Heijn instituted a prohibition on trade in South-African flowers.

By 1988, a German magazine published a list of products to boycott. This list included proteas and various other well-known South-African products with trade names such as Outspan, Cape Fruit, Del Monte amongst others. This on top of the prohibition of exports to the U.S.A. was causing problems for local farmers resulting in a certain amount of dumping on the Aalsmeer market. This resulted in a lowering of price, something which was rather detrimental in the long term. But the industry remained optimistic and saw a good future in protea farming, particularly since farmers were becoming more professional, getting away from veld harvesting and handling and packing were improving.

The Agricultural Produce Export Act, 1959, had been promulgated in order to establish, ensure, and maintain a high standard of normal agricultural export produce, mostly foodstuff, but also for flowers. Flowers are defined in the Regulations as 'the sexual reproductive parts of plants'. The legislators' purpose was to ensure that only flowers of good quality were sent to overseas markets, at the time most certainly only thinking of fresh-cut flowers. The use of natural dried floral materials for ornamental purposes was unknown at that time. Inspection standards for various types of cut flowers were laid down, at a later stage also for proteas, in cooperation with and welcomed by exporters.

When it was thought that dried proteas should be included in the standards, objections were raised. It was Walter Middelmann of Honingklip Dryflowers, representing the industry, who took it upon himself to prove to the department that no standards could be established since it was mostly the leftover flowers, bent or short stems, asymmetrical or deformed blooms etc which were used for drying. There was also discussion on how to define dried flowers.

The upshot of that was that a load of flowers was taken to Kirstenbosch where they were dried in an oven equipped with an air extractor fan at a temperature of 40°C for six days. Fourteen samples were tested and the average moisture content after six days was between 6.25 and 10.88 per cent moisture. The report stated that these results conformed to what is accepted in the pharmaceutical trade as the 'air-dried state', being between 10 and 12 per cent. It was found that 15 per cent would therefore be a reasonable standard. Then, at last, came a telex message from the department: *'Only 'flowers' with a moisture content of more than 15 percent will be inspected in future as published in Government Notice R2489 of 6/11/87.* Finally the department had given in and agreed that dried flowers need not be inspected. It was a long battle, but the industry won!

In 1988, a series of ten cultivars was presented with much fanfare to the S.A. Nature Foundation for multiplication for the industry to be done on their farm 'Protea Heights'. These were mainly from the genus protea as well as one *Leucadendron* selection. These cultivars had been selected and bred from 1973, some in cooperation with private owners.

There was progress all around. Before long we heard that Jimmy and Carol Yeats took on a partner and opened a fresh and dried wild-flower operation in Stanford. Jimmy and Carol had for years been farming at their farm Shamrock near Hermanus and were well versed in both fresh and dried operations. They concentrated on bouquet making. They later bought Witvoetskloof also in that area. The business did not make it in spite of a promising start. It was bought over by Pearl and Clive Gillman who eventually sold the operation to Bergflora.

A great event took place in Cape Town. Flora 88 followed in the footsteps of the highly successful Flora 83 SAPPEX had a small information stand at the event. A recently discovered new Mimetes, still unnamed at the time, would be on show for the first time. Also to be shown was the rare Marsh rose. The Mimetes was subsequently named *Mimetes chrysanthus* by John Rourke of the Compton Herbarium at Kirstenbosch. John was, that year, the recipient of the Compton Prize for the best publication in the *Journal of S.A. Botany*.

Penny Mustart, at the time a Masters student at the Ecolab of UCT's botany department, wrote an article for *Farmer's Weekly* in which she stated that 'understanding Fynbos is essential for wildflower picking and veld management is essential for its long-term maintenance'. All such articles were of course taken up by SAPPEX news and distributed to the members. Penny later joined the Institute for Plant Conservation at UCT, headed by Richard Cowling, which was set up from a grant by Leslie Hill. At one stage, Penny was a freelance tour operator leading botanical tours to Namaqualand. She is a specialist botanist, concentrating much of her work on the highly endangered cedars of the Cedarberg. These days she serves on the Editorial Committee of Veld and Flora.

In spite of the U.S.A.'s 'comprehensive anti-apartheid act' Phil Parvin of Hawaii was able to travel to South Africa, bringing with him the latest cultivars developed in Hawaii, New Zealand, and Australia for testing, in exchange for South-African material, in the interest of world horticulture. He brought with him twenty-six selections from these regions consisting of no less than 265 cuttings.

In October the same year, a New Zealander wrote to a South African supplier that sanctions against South Africa were in force *'I would not risk trying to pay you direct as I am already under surveillance by the local customs office. They are now opening every bit of mail, even the Veld and Flora newsletter and Bulletin'*.

The anti-apartheid row hit the Chelsea Flower Show. The organisers refused to withdraw their invitation to South Africa, *'we are a non-political charity and as such we are interested in our members and visitors seeing the best of the world's flora'* but some other exhibitors pulled out in protest. It was the last time that the Department of Foreign

Affairs participated. In 1987, they were asked to withdraw. It was acceptable for South Africa to participate, but not on government level and therefore the task was taken over by the National Botanic Gardens, Kirstenbosch, who walked away with a gold medal. It was not all quite new for Kirstenbosch because they had in the past coordinated flowers for the event, with flowers from the Gardens as well as from farmers and other botanic gardens in the province.

The Convention on Trade in Endangered Species (CITES), commonly called the Washington Agreement, came into being in 1987 and caused quite a stir. At first it was thought that all proteas would resort under this convention, but luckily this was not so. To be safe, the industry requested the Department of Nature and Environmental Conservation to issue a blanket certificate for plants from the Cape Floral Kingdom, provided exporters provided a pricelist and/or catalogue showing what species were exported. For items that were on the international list, like Aloe, a permit would have to be obtained, although it was *Aloe Vera* that was placed on the international CITES list and the South African *Aloe ferox* was exempt. For each export of *Aloe ferox*, the department now had to issue such certificates. The industry put a declaration on their pricelist that they observe this agreement.

Mr. Johann Eksteen of Elsenburg Agricultural College, who had been asked to do a survey of the protea industry, identified independent harvesters as the people most likely to strip the veld for quick gains, without worrying about flowers for the future. Again, this led to a lot of debate, but years later, in spite of discussion on all levels, no solution. It was impossible to control because absent landowners who did not really understand the value of diversity, were quite happy to earn a few bucks from the pickers, who in turn could always find an alternative property to harvest. There was, however, little or no proof of indiscriminate picking, so nothing could be done.

By 1987, SAPPEX gave attention to renewing the brochure, which would now also include some dried material, although the main emphasis was on fresh flowers. In the first brochure there was a recommendation that 'after a period of three weeks most products can be used as dried material'. I am guessing, and assume that the fresh flower exporters used this to encourage the florists to buy more material. However, it met with disapproval by the dried flower exporters who were developing really good-looking material and who felt that flowers just left in a bucket to dry were 'dead' flowers, and not nearly as good looking as those the dried flower trade was producing.

In 1987, Sharon von Broembsen received the Perishable Cargo Award for the best contribution to the Protea Industry for her work on diseases on proteas. Although the industry could not register any chemicals for prevention of disease, there were poisons registered for ornamental plants that proved to be quite useful. Again it showed that cultivation was the way to farm fresh proteas so that spraying against disease could take place.

Gerd von Mansberg of Indo-Atlantic Air Cargo, with Sharon von Broembsen receiving the Industry Award. With them is Anne Gray outgoing Chairman of SAPPEX. (Landbouweekblad 3 Julie 1987)

At Tygerhoek Experimental Farm, where the Protea Genebank of living plants was established, a Farmers Day was held under the theme, 'Market-orientated protea cut flower production'. Mr Tiel Bluhm, Director of Florimex South Africa in Johannesburg was invited to talk about current trends in floriculture marketing abroad. He rapped producers on the knuckles for not paying sufficient attention to detail in the pack shed and stated that the days of the Hobby Farmer were over.

Unless growers were prepared to be professional and do proper and consistent grading and packing, it would only backfire on them as they would not be considered sufficiently reliable to supply the exporters. Florimex, an international organisation buying and selling throughout the world certainly understood markets and marketing and we will be hearing more about Mr Bluhm later. Prof. Eloff of Kirstenbosch, another speaker at the Farmers Day, warned farmers to look after their plant resources as a lot of species were facing extinction due to urban sprawl, farming practices, and alien invasive plants.

The first snippets regarding a breakthrough in tissue culture for certain proteas were reported. This turned out to be premature and it is not yet really useful for the industry.

On a somewhat different note, SAPPEX News reported for the first time in 1987 on a regular cricket match between Exporters and Freight Agents played in Somerset West. It seems that Willem ('Fly Trap') Verhoogt was the hero for the Exporters with two excellent catches on the boundary, while Winston Odendaal as captain inspired the team. The freight agents lost by a mere eighteen runs. Score over the last five years, Exporters 4: Freight Agents 1. I wonder if they ever still play.

Dave Richardson and Richard Cowling at Fynbos Forum 2000 (LtR)

The Fynbos Biome Project that had its origins in about 1979 initiated a research program to study aspects of flower resources and natural vegetation of the Agulhas Plain. The initial stage of the research comprised a description and analyses of the region's flora and vegetation. The next phase was aimed at studying seed bank dynamics, germination ecology, and demographic responses to fire regime of selected species, studying those species that were being harvested or those that may have commercial potential. The aim was to stimulate research that is both practically relevant to farmers and conservationists and of sufficient significance to make an impact on the understanding of Fynbos plant community structure and function. SAPPEX donated

R5,000 towards this project. Here we first heard the name of Richard Cowling, who was later to be internationally acclaimed for his studies in the Fynbos, Succulent Karoo and later still the plant life of Southern Cape and Baviaanskloof — more about which is discussed later. Richard edited the book 'Ecology of Fynbos', which summarised ten years of research on Fynbos. He later brought out a magnificent coffee table edition with Dave Richardson as co-author.

A one-day symposium on the Wildflower Resource, organised by the Working Group of Commercial Wildflower Resources, funded by the Foundation for Research and Development, was held at Houw Hoek in 1988. Session one dealt with *Commercial Perspectives*, harvesting, commercial cultivation, and economic motivation. A talk had been prepared by Barrie Gibson of Mountain Range Flora, Ian Bell of Groot Hagelkraal and Walter Middelmann of Honingklip Dryflowers. The second session dealt with *Research Perspectives* such as veld resources, cultivation and cultivar development, and diseases and insect threats, while session three dealt with *Conservation Perspectives*, resources, and legislation, followed by a field trip to Honingklip Farm. An excellent report of the day's proceedings was published in Veld and Flora of June 1989. The need for improved communication between agencies representing commerce, research, and conservation was highlighted. The importance of this symposium could not have been foreseen, because eventually the Fynbos Biome Project and the Wildflower Resource Working Group gave rise to the highly successful annual Fynbos Forum conferences, held over four days, which now has over 300 participants.

In March 1988, SAPPEX visited the Eastern Cape growers. The Cape Town visitors stayed over at the Gamtoos Ferry Hotel on the Gamtoos River. I remember that Robert was called upon to rescue Alan and Renate Parsley who had got stuck somewhere in the Langkloof with car troubles. He made it back in time for a cocktail party that was held at the Thornhill Country Club where people could get to know each other informally. The next day everyone met at Kingsview Farm belonging to Nokie v.d. Burgh where the pack shed was converted to a lecture hall. Huge thunderclaps interrupted two of the speakers. After lunch, a practical pruning demonstration was to take place. Unfortunately, the rain had put paid to that, but with his usual practical mindedness, Nokie chopped down a large Repens 'tree' and Gerhard Malan could carry on with the demonstration. The visit was hugely successful and SAPPEX gained a good number of new members.

*Dr. Cobus Coetzee, ARC; Mr. Willem Verhoogt,
Berglora and Mrs. Renate Parsley, Parsley's Cape Seeds,
lecturing at Eastern Cape meeting, 5 March 1988*

Farming and conservation was receiving more attention and a very nice article in *Farmers Weekly* on Ian Bell of Hagelkraal near Pearly Beach emphasised his battle with invader plants. Subsequently, many years later Ian sold the farm, and now it is the preferred site for the next nuclear power station, and that in an area which has the highest endemism of plants in the threatened Renosterveld-type Fynbos of the Agulhas Plain!

In 1988, another case of a scarce plant being picked/stolen to near extinction, reached the press. This time the rare localised *Erica pillansii* of Kleinmond was decimated by illegal harvesting. It would be interesting to see the area now to check how it has recovered.

The demand for Everlastings for the dried flower industry probably reached its peak around this time. Particularly *Helichrysum vestitum* (now *Syncarpa vestita*) called Capblumen in the trade, was in danger of being over-harvested and the industry embarked on research on the possibility to establish commercial plantations. This never really came off the ground. Gerhard Malan who led the research found that of about 250 different species of wildflowers utilised in the Cape Floral Kingdom only twenty-eight kinds of Everlastings were being utilized by the dried flower trade, of which the Capblumen became known as white gold. It was not uncommon to see lines and lines of Everlastings bunches hung up to dry under the roof of many a producer and exporter. It was not easy to hang these, arms outstretched, under the roof in the heat of summer. Prices continued to increase and competition for sufficient product was quite fierce. The reason for this demand was the production of Dutch dried flower bouquets that incorporated Everlastings in preference to their own strawflowers that dropped their heads easily. Another advantage to using the Capblumen was that they could be dyed to match the bouquet colours. In the U.S.A. one of the biggest importers, having seen the popularity of the Dutch style bouquets, was importing and dyeing Everlastings by the container load in many colours for distribution. Eventually the product outpriced itself, and simultaneously the Dutch dried bouquets lost favour due to breakage — the soft garden flowers that they dried were not nearly as hardy as the South-African and Australia proteaceae and related greens, or the Californian spiral eucalyptus, broom-bloom, and other dried products. Now, in 2010, the Everlastings have become a minor sales item, and *S. vestita* is once again growing prolifically on the mountains. Farmers on the opposite side of our valley think it snows on our mountain in summer!

Both Ruth Middelmann and Marie Vogts turned 80 in 1988 and both got a write-up in SAPPEX News. They were certainly two well-known and well-loved personalities in the Fynbos arena.

In 1989, SAPPEX celebrated its twenty-fifth Anniversary with Maryke Middelmann at the helm.

Almost as an adjunct to his nursery work, Gert Brits had started to experiment with proteas as pot plants. During his many travels he had seen some potted proteas particularly in frost areas, where their owners could move them to a protected place to overwinter. By 1989, the first article appeared on his successes with **Leucospermum** and **Leucadendrons**, as well as with **Serruria**. Pot plants could be marketed within six to eight months after harvesting of cuttings. Especially the smaller varieties make stunning pot plants.

The Agricultural Research Council's Fynbos Unit (the last of many name changes) published a full colour A4-size booklet on its Fynbos work. This booklet was partially sponsored by the IPA and SAPPEX. By then the ARC in cooperation with the industry had already published its first cultivar poster depicting fifty-six species. Kleintjie Meynhardt initiated the ARC Fynbos Research Liaison Meeting, of which the first one was held on 14 April 1989. Scientists from Universities, the ARC, and industry sat around the table to thrash out a project plan for future research so as to avoid overlapping and duplication.

Regular items still appeared in *Farmers Weekly/Landbouweekblad* (its Afrikaans equivalent), for instance:

- January: Profiting from Proteas: A success story in the Winterskloof Valley near Hilton just outside Pietermaritzburg, where a 2-ha smallholding has become a model protea farm, with 100 different species and 20,000 plants, including the Waratah. According to the article, the word 'Waratah' means 'seen from afar' in the Aboriginal language — very apt, because the striking scarlet colour is visible from quite a distance.

(In the Western Cape, it is not a good idea to grow this species. We found that after growing them successfully for about twenty years or so, they are now starting to spread from their original planting area. So, here is another Australian species that, after an initial 'safe' period, can become a problem plant and invade the natural Fynbos.)

- May: Dream crop: Waratah. An article was published about Waratah being grown in northern Transvaal near Haenertsburg.

According to this article, the plant can grow very tall, and can reach an age of eighty-five years.
- July: Wild Flowers. In this article, written by Nature and Environmental Conservation, reference is made to a survey done by Melanie Simpson of Jonkershoek, who found that about 300 species of wild flowers were utilised by the industry in either fresh or dried form. A considerable portion of the trade consists not only of proteas, but of Everlastings, Brunia, Restios, Ericas, and other so-called 'greens'. The article again highlighted the problems arising from picking habits of independent harvesting teams working on land rented from absent landowners. Harvesting of 50 per cent of plant material or less was recommended. Various research institutions were looking at veld harvesting practices, amongst which, the Department of Nature and Environmental conservation at Jonkershoek, the Dept of Agriculture, Agricultural Research Council, and the University of Cape Town, the latter with Penny Mustart are doing wonderful work. Information was regularly fed to SAPPEX members. Research pointed to the fact that veld that had reached an age of thirty years or more, was no longer productive and fire would be necessary for regeneration. Farmers were advised not to harvest from the veld for one year before a planned burn in order that there should be a plentiful seed sources available for regeneration.
- November: Market asks for more Everlastings: Unfortunately, because of the high demand, producers pushed the prices beyond what importers and smaller wholesale buyers were prepared to pay, and eventually this 'greed' pushed the items out of the market. These days Everlastings are sold in small quantities, whereas up to the late 1980s the veld was probably over-picked. Fortunately there were sufficient seed stores to regenerate, particularly after fire.

(On our own farm, we felled a pine forest planted by Walter in the 1950s and after the final clean up in the mid-1990s and a subsequent fire to kill the last of the pine seedlings, Everlastings came up by their thousands.)

- November: Scarce Pincushion rediscovered: The find was a yellow *Leucospermum patersonii* found on the Jackson's farm near Stanford. Apart from this new variant, there are only three yellow pincushions of which one, Yellow Bird, a variety of *L. cordifolium*, is now extinct in nature. The Jackson's donated plant material to the ARC and also started propagation from cuttings to multiply this scarce pincushion (sunburst). Yellow Bird, a semi-prostrate species, was eventually crossed with *L. patersonii* (yellow) to produce an upright yellow pincushion with the name 'High Gold', which has become very popular.

The alien invasive plants that caused such problems in various regions of the Cape Floral Kingdom were of great concern to Fynbos farmers. Barrie Gibson on behalf of the industry wrote to Mr. G.S. Bosch of the Western Cape Agricultural Union asking what the position of the industry was in terms of agricultural recognition, and whether any assistance could be given to Fynbos farmers to clear their land. The WCAU wrote back that Fynbos Farming was not recognised by the Department of Agriculture, although they admitted that cultivated proteas should be recognised as an agricultural pursuit. All they could suggest was that the industry makes use of the biological control available from the department, as chemical control would not be an option in the Fynbos. The steep slopes involved were likely to cause pollution of soils and water courses.

A decision was made by the department to move the Genebank from Riviersonderend to the grounds of the Elsenburg Agricultural College, outside Stellenbosch; a move that took until 1992 to accomplish. All the accessions and mother plants had to be carefully made into cuttings for re-establishing at Elsenburg. It was quite a task.

SAPPEX was informed by the Directorate of Agricultural Product Standards that the issuing of Phytosanitary Certificates for dried flowers would no longer be carried out because quality inspections no longer took place. After some discussion on the subject, they agreed to do inspection for diseases and insects for which they would issue certificates, but they could not inspect whether products were up to the required quality standards.

The fresh flower industry was experiencing problems with inspection of particularly *P. magnifica*. The department and the industry solved the problem by commissioning a wall chart, depicting the good, the bad and the ugly and everything in between. This wall chart would be used by Magnifica producers as well as the inspectorate and served to clear up misunderstandings and incorrect interpretation of the regulations.

The University of Cape Town announced the Protea Atlas Project headed by Tony Rebelo. Tony was to harness botanists and enthusiasts throughout the Fynbos Biome to go on excursions and report on proteaceae locations and densities. He set about coding and references that could be correlated into a database. It was going to be quite something to get this off the ground. The Protea Atlas Project was to run from the Kirstenbosch offices.

The South-African industry recorded a turnover of R29 million in 1989. The ARC announced that research on Fynbos would in future have to be paid for and this was to lead to difficulties in more ways than one, eventually causing a breakdown of industry unity — but more about that later.

Meanwhile the South African Fern business had starting to develop nicely, mainly in the Tsitsikama area, where it was initiated from the harvesting of Ruhmora and Coral fern in forestry land. The ARC suggested that, seeing it was just another form of wild harvesting, it should fall under SAPPEX so that the Department could take this new product under its research wing. As a result of this new development, the first ever meeting was held in that area in 1989. Because the response was so good, we had to move the venue from Forest Ferns, to the hotel where we were staying overnight, which had bigger facilities. I seem to recollect that it was the first meeting where SAPPEX asked its members not to smoke in the lecture hall. These days it is hard to comprehend how people used to smoke everywhere, even in cinemas.

SAPPEX found it was time to look at participating in Trade Shows. At the time it was quite acceptable to have a little black and white leaflet, giving some details on the association. Many years later, for Washington DC we, together with the ARC, produced a lovely cut-out full colour leaflet.

International Developments

Australia

From Australia came a document that outlined their proposed packaging and standards for proteaceae in Australia. It was a well-designed document that should have been copied in different producing areas, with minor amendments for local conditions to bring standards to a uniformity that, in my opinion, would have made sense internationally. Unfortunately, it did not happen.

In February 1986, the first journal of the Australian Protea Growers Association (APGA) saw the light. The APGA had branches in each state and they gathered information from the branches for publication, as well as publishing freely from other associations around the world on a reciprocal basis. This excerpt was from APGA News. 'We had all hoped for good prices and large orders. We received better prices, but due to the economic position and political climate which dampened air travel, quite a few airlines cancelled flights, or took on less freight to cut on fuel usage to cut costs. What happened in the end was the significant drop in freight space and a fight between exporters to obtain space.'

A dried flower export business developed in Australia. The firm Australian Flower Exports (Pty) Ltd. produced two very nice catalogues in 1987, one for dried material and one for fresh flowers.

James and Elizabeth Wood of Protea Pride, Busselton (ex South Africans) joined the big league of Western Australian business and were finalists in the State Industry and Export Achievement Awards, a mere three years after arriving in Australia from South Africa. In the same year, the Protea Growers Association staged its inaugural Protea Flower Festival showing off the flowers of Western Australia.

All the information about international production of proteas was becoming a topic, which prompted the Consul (Agricultural) in Beverley Hills, California, to write a special report on Proteas in Hawaii, September 1987. This report mentioned the 'royal' names coined by Hawaii, namely the King (Cynaroides), Queen (Magnifica), and also Prince (Compacta), Princess (Grandiceps), Duchess (Eximia), and Jester (Obtusifolia) The

Neriifolia types were named Minks. These names were first mentioned by Phil Parvin who spoke on the Royal Family of Proteas at the second IPA Conference that was held in 1983 in the U.S.A., according to those excellent little books that Peter Mathews produced after each conference, entitled 'Growing and Marketing of Proteas'. Each of these books contained a full transcript of talks and discussions, reports, and list of participants. These days the IPA provides an Abstract book and the International Society for Horticultural Science approves publication of an ACTA of the scientific papers presented at the IPA Conferences.

IPA

The IPA had embarked on producing, in conjunction with South Africa, a brochure showing worldwide growing months and depicting both the Australian and South-African genera. One of the aims was to educate the buyers, and for that reason it was hoped to get some uniformity in names. From California came their brochure regarding what varieties were grown there and under what name. Dr. Peter Sacks was not very optimistic about being able to convince people around the world to use standard common names as most names were well entrenched in their own growing areas. The table below clearly demonstrates the problem with different common names used in different countries for the same product as well as same names for different products:

Same product different names	
South Africa	**Australia**
Klaas Louw	Morrison Nitens
Bunny tails	Pussy Tails
Spidergum	Bushy Yate Nut
Willow eucalyptus	Bookleaf
Different product, same names	
Lambs Tails	Lambs Tails

With common names even different in regions within South Africa, it is hardly surprising that this problem is not an easy one to overcome.

Then we heard that Peter Mathews had had a nasty fall in January when a ladder collapsed under him while he was getting off the roof. He broke and crushed a vertebra and was laid up till March, but one of the first things he tackled when he was allowed to do a bit of work daily, was to continue efforts on the IPA brochure.

IPA Protea Brochure 1988

The next IPA Conference was scheduled for Auckland, New Zealand, in 1987. South Africans in particular had to overcome Australia's very strict visa problems. Walter Middelmann had taken on the mammoth task of coordinating photographic material from all over the protea producing world for the IPA brochure. He gave a full report on how at last, after two years of effort, the brochure was ready for distribution. Twelve thousand copies had been printed of which half were to be taken by South-African exporters and the balance would be available for other countries. No additional funding was required from the IPA because the South-African conference had made sufficient profit so that all costs were met from the South-African held IPA account, with some left for other IPA use in the future.

By this time certain procedures had already become entrenched in the IPA, like Board member's country reports, the research reports from each country, and the research committee (who interacted strongly with the IPWG) report. At the conference, scientists submitted research proposals for funding, which were first handed to the IPWG for recommendation to the IPA research committee and ratification by the Board. The IPA journal was a valuable document as it published articles from various associations as well as research results from all over the world, thereby increasing knowledge on the South-African and Australian genera of proteaceae.

After Peter Mathews' retirement, the reigns were handed over into the capable hands of Dame Joyce Daws, with David Mathews as Secretary. With a succession of dedicated Journal Editors, the IPA was in good hands.

In *Western Farmer* (W. Australia) of 1988, there was an article on Maggie and Mark Edmonds and their protea operation. Maggie would eventually become the IPA President for the 1991 Perth Conference.

The next conference was scheduled to take place in San Diego, California in 1989 with Dennis Perry as President. At the conference, Dr Marie Vogts was unanimously voted an Honorary Member of the IPA. Sharon von Broembsen's book 'Protea Diseases', published by the IPA, was launched there. It was a real first and although it consisted of only forty-three pages, it was packed with information on root, shoot, stem, flower, and leaf diseases of proteas. There was even a chapter on formulating disease management and another on sanitation pruning. The illustrations made this book a real boon for protea farmers around the world. Although Sharon was born in the U.S.A., we considered her a true South African and it was thus with pride that South Africa could say that two international publications originated in our country.

As always, Walter, the South African Board representative, firstly presented a report on the South African Industry, and then afterwards wrote an excellent summary of the conference, which again had run concurrently with the IPWG symposium. The IPA brochure launched at the previous conference had been so successful that it entered into a second print, totalling 20,000 copies. The Board received a request to produce a wall chart, similar to the Dutch Flower style wall chart.

The conference was attended by seventy full delegates, and numerous daily visitors from the Escondido area. A pre-conference tour of Hawaii (Maui) and a post conference tour up the Coast as far as San Francisco was organised for overseas visitors. After all, theory is one thing, but to really see how things are done, that is what sticks in the farmer's mind. Walter and Ruth, together with Renate and Alan Parsley did their own 'post-conference' in order to visit the Arboretum at Santa Cruz where they were to first lecture and later lead a tour of the Arboretum's Southern Africa collection.

Of course, reports from the Conference filtered through to South Africa and quite often the press got it wrong. There was, for instance, a report in the Cape Times that New Zealand had overtaken South Africa as the world's number one producer of proteas, together with such crazy statement as 'the Kiwis are coining it by growing proteas in hothouses'. As Walter wrote in response: *'In terms of flower production, only the Leucadendron cultivar 'Safari Sunset' is of economic importance there. As to 'proteas in hothouses, they would all just die!'*

International Protea Working Group

This organisation grew in strength after the South-African symposium, which was held concurrently with the IPA Conference and they even had an application for membership from a scientist in Germany.

Gert Brits who was the person responsible for the Protea Register (South African genera) wrote to IPA representatives, asking them to forward the name of a contact person to assist with registration from their countries so that the first list could be drawn up. Accompanying this letter were guidelines for registration of cultivars. This followed a decision of the International Union for the Protection of New Plant Varieties (UPOV) to compile guidelines providing a standard for testing and examination of new protea cultivars for plant breeder's rights. These guidelines were to be published by UPOV for the genera *Leucadendron, Leucospermum,* and *Protea*.

Protea News recorded the death of Dr. John Salinger of New Zealand in February 1987. After his retirement in 1984 he wrote the popular and informative book 'Commercial Flower Growing and Marketing' specifically produced for New Zealand growers, based on his long experience in horticulture.

Shortly after the New Zealand conference in 1987, Phil Parvin lobbied Dr. Furuta of the University of California to join the Protea Working Group and to organise the IPWG symposium in 1989. Via Dr. Richard Criley, the Vice President for North America (ISHS Section Ornamentals) the ISHS Executive Committee was approached, who duly approved the request.

U.S.A.

For the first time in 1986 mention was made in Pacific Horticulture of using proteas as dried flowers. These were not treated in any way, just naturally air dried, although the picture featured with the article actually shows a *P repens* with the seeds removed. Dried flowers were never a commercial success in Hawaii, California, or anywhere else where they were cultivated, as it made them far too expensive to market. Only wild harvested species in Australia and South Africa could be produced cheaply enough to be commercially viable. There were hobbyists who created interesting trinkets like dried protea flower dolls to use as Christmas decorations, pictures, and so forth, but never on the same scale as South Africa who exported this as a commodity in bulk boxes all over the world.

Probably the most widely read magazine in the Western World, *TIME*, for the very first time mentioned proteas in an article on 'Sunny days for Flower Sales' in its issue of 17 February 1986. The article itself described the increasing interest in the use of cut flowers in the U.S.A. where retail sales had increased from US$2.9 billion in 1981 to $3.7 billion in 1985. For us, the most relevant part of the article read:

> *Demand for exotic foreign flowers is also booming. They accounted for an estimated $240 million in sales during 1985. The most popular source is the Netherlands, which exports about $100 million worth of tulips, freesias, lilies, alstromerias and other varieties to the U.S. each year. Colombia and Israel are also major suppliers. From Hawaii and the Caribbean come tropical strains such as orchids, heliconias, proteas and banana flowers.*

Hawaii was really good at promotion of proteaceae as tropicals! People in the trade of course knew better, and knew the source of these flowers to be South Africa and Australia.

Dennis Perry, Greenhouse Manager, stated that he became interested in proteas as a result of a lecture by one of his teachers at California Polytechnic University at Pomona, in which proteas were briefly mentioned during a lecture on alternative crops. This piqued his interest, but there was no information until his father brought seed back from South Africa in 1979 with all kinds of information, which he reckoned did not help him at all. After several years of experimentation, he became a real pioneer in California growing proteas for cut flowers and in pots.

A list on what proteas are suitable for what soil types published by Dennis warrants copying here:

Proteas for clay soils

Protea cynaroides
Protea burchelli
Protea obtusifolia
Protea mundi
Protea repens
Banksia speciosa
Leucadendron eucalyptifolium
Leucadendron argenteum
Leucadendron salicifolium

Proteas for 'garden watering'

Protea eximia
Protea cynaroides
Protea lepidocarpodendron
Protea mundi
Leucadendron salicifolium
Leucadendron argenteum
Banksia ericifolia

Protea for alkaline soils

Protea obtusifolia
Protea repens
Protea neriifolia
Protea eximia
Protea cynaroides
Protea laurifolia
Leucadendron argenteum
Leucadendron salignum
Leucadendron uliginosum
Leucospermum patersonii
Banksia prionotes

Protea for frost tolerance

(below 20°F)
Protea repens
Protea laurifoliua
Protea grandiceps
Protea magnifica
Protea neriifolia
Protea eximia
Protea burchelli
Protea cynaroides
Leucadendron eucalyptifolium
Leucadendron salignum
Leucadendron salicifolium

Protea for patio tubs

Protea cynaroides
Protea magnifica
Protea grandiceps
Leucospermum tottum
Leucospermum cordifolium
Leucadendron salignum
Leucadendron gandogeri
Banksia menziesii

Protea for drought tolerance

Protea laurifolia
Protea lorifolia
Leucospermum reflexum
Leucospermum vestitum

In 1988, the marketing department of Advanced Floral Concepts of Cedar Falls, Iowa, picked up on the term 'Pink Mink'. This was featured in their *Design for Profit* issue Vo. 11. No. 4, Nov. 1988. Developed in 1987 to increase sales of *Rainbow Cork* they manufactured, they promoted the use of the slogan 'a Mink for You' and produced a florist package containing:

> Pink Mink merchandising tags
> Buttons (would you like a Mink?)
> Point of purchase counter signage
> Consumer information handouts
> How to do it (for design room use)

They also designed and produced cards and paper to add to bouquets and vases.

"Pink Mink" *P. neriifolia* used for major promotion brochure, USA

Hawaii

In one of Proteaflora's newsletters in 1988 it stated that *'Hawaiian Proteas are primarily grown on Maui in volcanic soil and in semi-tropical temperatures. They experience good growth but have fungal problems to contend with. Their greatest problem is Pincushion scab disease. Through cooperation between the University of Hawaii and the growers they have set up a plant breeding and selection program to ensure the availability of good stock. 'Hawaii Gold' was their first release. A further three releases are envisaged this year.'* On the Island of Hawaii the proteas were healthy looking. They were grown on old lava flows flattened by a bulldozer. They were drip irrigated and fed the major nutrients (NPK) as well as being subjected to a regular fungicide spray program. Phosphorus toxicity appeared to be a problem in Neriifolia. Most of the product from Hawaii went out in small gift packs, to the customer's door. The flower value was $10 as compared to the delivered cost at $60.' Leaders in the industry, just like the Mathews family of Proteaflora, have always travelled with open eyes and mind and shared information in order for their local protea industry to learn and improve.

Superb advertising leaflets doubling up as mail order forms were produced by Protea Gardens Maui, where complete bouquets were produced at fantastic prices, for instance almost US$30 for a bouquet of six proteas with a variety of foliage and that in 1987, or US$45 for an assortment with either a Cynaroides (King) or Magnifica (Queen) included. They also accepted orders for one bouquet per month for a year, for a price between $310 and $500.

One of the early Hawaii Growers was Colin Lennox, a retired Agriculture expert. On Cloud Bank Farm in Kula, Maui, he produced two hybrids. He was certainly a highly respected and very involved person, who earned himself an honorary doctorate of Science from the University of Hawaii. When he died in a motor car accident in 1989 his obituary in the Honolulu Advertiser was sent to the IPA.

New Zealand

Jack Harré published his very successful book 'Proteas' intended to help potential growers to establish proteas in cultivation. Thirty years of experience in growing, eight years of research, and careful observations on frequent trips to protea growing regions throughout the world resulted in a practical guide to propagation and growing proteas. Phil Parvin wrote the book review.

Ken Joyce of NZ was one of the keynote speakers at fourth Australian Protea Growers Association Conference, NSW on *'A model of good crop management and successful marketing'*. Ken had established proteas on his 10-acre property. He was speaking from experience in marketing to Japan and the U.S.A. West Coast.

Canary Islands

The first report on protea cultivation on the Canary Islands came from Ms. Pilkington, an ex South African, who bought a few protea seeds in 1986. She reported that the few seeds she had experimented with were doing well and wrote that she would like to try and grow proteas on a large scale in the Canary Islands.

Russia

Although we are yet to hear of production in Russia, it is interesting to note that in 1989 the first ever Miss USSR was selected from 500 regional contests. Yes, so what does this have to do with proteas? Ah, according to the photo in *Time* magazine of 5 June that year, Miss USSR, Yulia Sukhanova, aged seventeen received a fresh flower bouquet containing a ***Protea repens*** and what looks suspiciously like a ***Leucadendron*** of some nature. It could be a Safari Sunset. Unfortunately, the picture I have is photocopy of rather poor quality. But it was nice to see that proteas had reached Moscow!

Israel

Israel took over as the largest producer of Safari Sunset. Unfortunately, the leaflet of recommendations is undated. Fuga Nurseries, who published this leaflet had started production in the early 1980s and had reached the level where field trials in the production of grafted Safari Sunset plants were receiving attention. The reason for grafting was to adapt the plant to unsuitable soils, with Israel having neutral soils of about pH 7. They recommended a plant density of 700 plants per 1000m^2.

A bilateral agreement had been reached between South African and Israel. On the strength of protea plantings there, this agreement could be strengthened. Students from Israel would be able to assist with research on subjects where there was a shortage of researchers, and South Africa would, against the background of political problems, benefit from proteas 'loosing' their South-African identity.

Japan

In Japan meanwhile, a further publication on Fynbos, with emphasis on *Leucadendron*, appeared in *Garden Life*. This was arranged by Mr. Sugiyama, Manager of Flora International, and Chairman of the Japan Floriculture Organisation. He had called on Walter Middelmann to assist him, as Walter was, at that time, the only commercial person to undertake regular trips to Japan. An interesting leaflet, only in Japanese, shows Proteas, Leucadendrons, and Everlastings in nature as well as in their dried form. The pictures came by courtesy of Walter, who also supplied pictures of some of the work by Rae Steyn of Kleinmond, in the form of dried flower pictures, posies, button holes, and even a bridal bouquet of dried blushing brides (*Serruria florida*).

Industry Developments 1990 - 1994
South Africa

In 1990, the South African botanic fraternity, joined by many people worldwide, mourned the death of Brian Rycroft. He was responsible for a period of major development and expansion of the National Botanic Gardens. Under his guidance and vision, Kirstenbosch in particular became a botanic garden of world renown. He was awarded the Decoration for Meritorious Service to the country in 1980. In his final tribute, Prof. Jackson said: 'Above Pearson's grave in this garden you will read, *if Ye seek his monument, look around.* You can't do this for Brian Rycroft — you would have to travel to the four corners of the country to do that'.

Fortunately, the cordial relationship that existed between Kirstenbosch and SAPPEX was continued with the appointment of Brian Huntley, who like his predecessor became a personal friend of the Middelmanns. Kirstenbosch of course was also doing a lot of research, developing restios and other products as garden plants. Their annual plant sale is legendary. It still amazes me that Ericas were never successfully raised as pot plants in South Africa in spite of Europeans having done it for many years and in spite of efforts by the Department of Horticultural Science, University of Natal, Pietermaritzburg. Landscapers recognised the potential of these plants for landscaping and initiated research by horticultural students on this aspect. One research paper on the subject reached us via Parks and Grounds.

At a Farmers Day held in February 1990, AGM James Wood from Busselton, Australia gave a landmark speech on priorities for protea growers. This talk was published in SAPPEX News and Roodeplaat Bulletin. We have had the pleasure of visiting him there a few times and were really impressed by what the family had achieved in a very short time.

At the same meeting Kleintjie Meynhardt announced that the newly formed ARC would approach research differently and that he expected that research on bulb plants could very well receive higher priority than proteas. He stated that the industry would have to pay for near market research, while long-term research would still be the responsibility of the ARC. He warned that a contribution of 10 per cent of the ARC's Fynbos research budget would have to be made by industry. He said that if farmers were to use 1 per cent of their income for research and development, it would be able to fund research to the tune of R1,800,000 double the budget at that time. It took a while for this to sink in and for the committee to take action.

In 1991, SAPPEX initiated a survey on the costs of establishing plantations under cultivation. Dr. Gerhard Malan of ARC was asked to obtain information from growers to correlate and come up with an average cost. By then there was such interest in protea growing because of the many articles being published in *Farmers Weekly* and *Landbouweekblad*, that the secretariat needed some facts and figures to impart to potential members and growers. Although of course not everyone participated, the first report made in March 1993 did bring some factors to light. The lack of information concerning the market status of individual products was cited as the most important factor for farmers to make a decision on what to plant. There was still a lack of knowledge on pruning and fertigation. The report listed the most popular items harvested and grown and gave exporters valuable information on what to expect to have to move in the future.

Also that year, with generous assistance from SAA, who supplied a free ticket plus 200 kg freight free of charge, and with help from the South African Embassy and Perishable Cargo Agents (PCA) in the form of Gert von Mansberg, the Protea Industry under the association's banner, had a stand at the Taipei Flower Show. Lea Liebenberg did the decoration of the stand, while Gerd von Mansberg of PCA manned the stand for the duration of the show. Robert Middelmann who was on a business trip in that region at the time, gave some moral assistance and volunteered some time. Other South Africans visiting the stand were Frans Gerber of Forest Ferns, Urs Jucker of Bergflora, and Paul v.d. Berg. The reason for attending the show was to test the market in an attempt to decrease dependence on Europe.

Devastating fires in Du Toitskloof and elsewhere resulted in 'Fire' being the theme for the 1990 AGM. Although a number of articles subsequently appeared in the *Farmers Weekly* on this subject, no action was taken by those in authority. It took at least another decade before the authorities started to take notice of fires in the Fynbos. One of the problems was that the authorities themselves had neglected preventative block burns and fire belts, increasing the danger of runaway fires. All farmers could do was try to minimise dangers on their own property and to take out public liability insurance. The Minister of Environment and Water Affairs was specially invited to hear our problems. He, however, reminded the industry that they should pay for research and development. In this talk he defined conservation of wild resources by stating that '*conservation demands that utilisation does not exceed the ability (of the veld) to recover,*' which would of course be in the interest of the industry.

Subsequently, an industry think-tank resulted in a research levy, based on the number of kilograms exported, so that SAPPEX could meet the demand by the ARC that the industry must pay for research. The first industry payment for research was made in 1992. Over the years, quite a lot of money was invested by SAPPEX, via its membership contributions, into research, mainly to the ARC, but also later to other institutions.

SAPPEX had become a member of Union Fleurs, an international organisation of Floral Wholesalers in order to use them for an EC tariff reduction application. To this end, Barrie attended a Union Fleurs meeting in Cambridge in 1992 to ask for a reduction in affiliation/membership fees, which were normally set at Deutsch Mark 3,200. After some difficulty, the membership fee was reduced to DEM 1,000 per annum. SAPPEX paid this amount for many years and this affiliation did not offer much in assisting SAPPEX to overcome trade barriers. Apart from that the volume of documents they generated was quite a nuisance to wade through and eventually SAPPEX resigned from the organisation.

Some of the articles that appeared in *Farmers Weekly* and *Landbouweekblad* over this period were:

- 'Variety Vital in Protea Market'. The new protea cultivars now available as rooted cuttings will enable producers to establish plantings quickly and easily.

- 'It Pays to Look after Fynbos'. There's money to be made from Fynbos, but it must be managed correctly. This article referred to harvesting from nature.
- 'Companion Crops for Bigger Protea Profits'. There's a lot of potential in growing other indigenous flowers in conjunction with proteas.
- 'Protea Knelpunte uitgeskakel' (Protea barrier overcome — improved plant material now available.
- 'Conservation and utilization go hand in hand'
- 'Item on dried flowers' (Louis Mostert)
- 'Smoke treatment boosts Fynbos'
- 'Boost for Wildflower Industry — Fynbos Genebank at Elsenburg'
- 'Australian Farmers of Proteaceae' (*Landbouweekblad*)
- 'Good Overseas Market for Ferns'
- 'Protea beat the drought — Proteas in the Transvaal'

The *Cape Times* published a picture of dried flowers decorating the St. George's cathedral for Red Cross day. At last the dried flower industry was getting some recognition!

By then, with financial assistance from a member of SAPPEX who preferred to remain anonymous, the ARC had established a cuttings house for rooting cuttings from mother stock. The intention was that these cuttings would be established by farmers as their own mother stock from which they could eventually establish a plantation. In many cases, the farmers merely used them as trial plants from which they sold some flowers once they got to that stage, and if successful, they ordered more of the same cultivars. Very few members had their own nurseries for rooting cuttings. In this way it took ages for the cultivars to be commercially established and the ARC came under fire for not being able to supply sufficient cuttings.

Management of the Fynbos Genebank was taken over by Gail Littlejohn. In an article in *Farmers Weekly* of September 1994 she stated that apart from proteas, collections had been made of Brunia, Erica, Retzia, as well as bulbous species, with over 1450 variants of more than 80 woody Fynbos species and over 1000 variants of more than 80 bulbous species. According to Gail, income generated from harvesting from 50 to 100 hectare of natural Fynbos was equivalent to that of one cultivated hectare.

In 1993, Dr. Kleintjie Meynhardt died. He had served at Roodeplaat for twenty-five years, and became the Deputy Director in 1968. He was badly missed by both his friends and colleagues. As far as the industry is concerned, he will probably be best remembered for being the inspiration behind two accomplishments in the protea world: the release of the first protea cultivar and, together with Phil Parvin of Hawaii, the formation of the International Society of Horticultural Science's (ISHS) International Protea Working Group (IPGW).

In 1993, SAPPEX published its first information leaflet for handing out to interested people at shows. This leaflet listed 9 dried flower exporters, 11 fresh flower exporters and 6 seed suppliers. It was just a little black and white folder A4 page, giving some background on the association with the logo.

In 1994, a protea farmer claimed R3.1 million from his neighbour who had allegedly started a fire without ensuring that he could control the extent of the fire, or be certain that he could extinguish same. The farmer claimed for loss of income on his proteas and repairs to fencing and cables. It was the first of a number of court cases resulting from fires spreading to neighbouring farms. In spite of many attempts, the industry was not successful in obtaining insurance for cultivated proteas, due to the high risk and the difficulty of measuring value. Underwriters were just not interested in such a strange and variable crop.

The airfreight subsidy on the Cape Town/Johannesburg leg that the industry had enjoyed up till then was drastically cut on 1 April 1990. This caused quite a stir amongst the exporters, but nothing could be done to rescue the situation and exporters had to bite the bullet. Fortunately, they had sufficient advance warning before the onset of the new season in October. Airfreight space during the export season was always problematic due to simultaneous fruit and fish exports from the Cape, but in time more airlines came directly to Cape Town, which brought some relief.

Willem Verhoogt, a leading exporter, warned farmers that they would have to take cost-saving measures to keep prices low and quality would have to be very high if proteas were to remain competitive on the European market. Better packaging methods would help to increase the volumetric mass, and so better utilise airfreight. Proteas were at that stage still banned by the U.S.A., due to the number of insect interceptions.

Conservationists had started to take notice of protea farming in and of the natural resource and were concerned about the implications for conservation. George Davis of the Stress Ecology Unit circularised the Wildflower Industry Working Group that he intended calling a workshop in 1992 to address sustainable veld management for the wildflower industry. Controversy ranged around the subject of wild flower picking. Conservationists argued that it could put many more species on the endangered list, while businessmen suggested that harvesting could be the saviour of the Fynbos. It was high time to get the role players around the table. Most farmers were, in fact, practising sustainable harvesting. The problem mainly lay with lessee harvesters who did not leave sufficient good-quality seed-bearing heads on the plants for future generations. The workshop on the utilisation of Wild Flowers on Natural Fynbos was held at the Brandfontein Private Nature Reserve. I attended this workshop and was introduced to conservation issues and conservation personalities like Charlie Boucher, Christo Marais, and others.

Pick 'n Pay, Captour, and WWF South Africa published a leaflet 'Guide to the Fynbos of the Western Cape' for distribution to the tourism industry. Tourist knowledge about our Fynbos was rather scant and such initiatives led to popularising proteas and evoked interest in buying them when tourists returned home.

A person on his way to international fame was young Richard Cowling, who already in 1987, a mere three years after completing his studies, won the CSIR President's Outstanding Young Scientist Award. Richard Cowling, then of the Botany Department at UCT, was one of the main role players in creating awareness of the species richness of the Agulhas Plain. A research project was initiated on the Agulhas Plain in January 1990. The goals of this landmark project stated, amongst others that: *The coastal forelands of the south-western Cape include a wide array of species-rich Fynbos communities with many hundreds of locally endemic species. These communities are severely threatened by alien invasive plants, resort development, injudicious use of fire and over exploitation of wildflowers. There is an urgent need to gain a predictive understanding of ecological processes in these communities and apply this knowledge for their wise management. Virtually nothing is known of the impacts of vegetation fragmentation or flower harvesting on community structure and processes. The challenges of the Agulhas Plain should stimulate research that is both practically relevant to farmers and conservationists.*

Professor Richard Cowling

A report was penned by Richard for the research section of UCT News entitled 'Farming Fynbos — Reconciling conservation with exploitation' that illustrated the species richness of the area and the challenges facing the wildflower industry. This initial project eventually, after many years and a lot of work and effort to get farmers on their side, led to the formation of the Agulhas National Park.

When UCT was granted an endowment to establish a Chair of Plant Conservation, Richard Cowling, was appointed professor of this new unit. I was very fortunate to attend his inaugural speech at UCT. It was probably also the one and only time that I saw him dressed in a tie and jacket. Richard to this day remains a leader in the field of conservation, having earned himself an Award by the Pew Charitable Trust in 1994, one of ten such awards that are given annually worldwide to individuals whose careers reflect a commitment to both scholarship and environmental action. He subsequently moved to Port Elizabeth where he immersed himself in the local vegetation and was one of the leading lights in having the Baviaanskloof declared a mega-reserve. But to this day he likes to keep his feet firmly on his surfboard!

With commercial and conservation research increasing, interesting data came to SAPPEX who regularly published snippets on veld management. We all knew that veld fire was necessary for renewal, but not quite how and why. Researchers found that germination was not only the result of burning to release seeds, followed by rain to wet the soil, but also that temperature differentiation had much to do with it, requiring about 20°C during the day and 10°C at night. Then came the NBI scientists, Dr. Neville Brown and Dr. Hannes de Lange, who discovered that smoke was an important primer for seed germination. To isolate the critical elements in smoke which stimulate the seeds was the real question, with several hundred substances in smoke, anyone of which could be the key! But it did not take long before Kirstenbosch released 'smoke primer' for hard-to-germinate species.

In 1991, Joan Pare, our incomparable Protea Ambassador extra-ordinaire, was attacked at her Constantia home. I thought that it might be nice for her to get a letter or note from the Queen to cheer her up, so I sent a short letter to Buckingham Palace enclosing the newspaper report of the attack. Instead of receiving a note from the Queen, Joan was invited to 'meet the Queen' at the preview of the next Chelsea show. Although I had not expected a reply, I have a letter addressed to me from Buckingham Palace 'I am commanded by The Queen . . .' signed by Susan Hussey, her Lady-in-Waiting. I doubt if Joan ever realised how the invitation to meet the Queen came about. Over the years Joan had built up a veritable collection of letters from Buckingham palace which, unfortunately, all got burnt in the Cape Peninsula fires of 2000 that destroyed her home. Fortunately, SAPPEX had copies of a number of them and these were sent to her daughter shortly after Joan's death.

In 1992, South Africa lost a true protea ambassador when Renate Parsley, one of the most energetic and enthusiastic people I knew, died in April. Having bought Ruth Middelmann's original proteaceae seed business, she expanded it and although a complete amateur, having qualified as a radiographer, became renowned for her pelargonium knowledge, to such an extent that she was invited to be on the British National Pelargonium Collection's stand at Chelsea. Strangely enough it was music that brought Renate into contact not only with Ruth, but also with her husband Alan, through the St. George's singers (of the St George's Cathedral in Cape Town). She served on the SAPPEX committee and was the driving force behind rejuvenating the Mini Competition. This competition was for promising foliage or flowers for export. Farmers could bring material to every SAPPEX meeting. The competition was renamed in her honour.

When Kirstenbosch announced in Veld & Flora of September 1992 that they also now had the cultivar *Leucospermum cordifolium* 'yellow bird' available as a welcome addition to the colour range of this species, Ruth Middelmann put the record straight: *'I would like to point out that the yellow form of Lsp. cordifolium was found by myself growing among the rocks above the Bot River on the side of the former Hermanus Road many years ago. I collected seeds and grew this form successfully on our farm "Honingklip" on the opposite side of the Bot River. I do maintain that the "yellow bird" should not be determined as a cultivar. It is a clear yellow form of the species Leucospermum cordifolium — in the locality where I found it, it was very plentiful.'* This yellow cordifolium no longer grows there, but has been preserved in the Fynbos Genebank in Stellenbosch, where it was crossed with *L. patersonii* to become the popular cultivar 'High Gold' (*Landbouweekblad* 20.3.92).

The Botanical Society, which had always sold Kirstenbosch's excess seed to their members worldwide, published a small leaflet on how to grow Cape Proteaceae. They also listed the commercial seed nurseries. They came under fire for wasting these seeds on gardeners who are not trained to propagate plants, rather than to supply commercial nurseries who were gearing up to market 'indigenous' sections and were looking to Kirstenbosch as being on the forefront of knowledge regarding suitable indigenous species for local gardens. As a result Kirstenbosch undertook to only provide quality seed through a network of garden shops located in all the regional botanical gardens. It was not in the NBI's mandate to be involved in the horticultural business.

At least one Farmers Day was held annually, more than often jointly organised by VOPRI and SAPPEX. These events were normally well attended. VOPRI started a Fynbos Plant Protection Section at Elsenburg, under the leadership of Mark Wright, who was later to move to Hawaii. In 1993, the ARC had used a stylised *Protea cynaroides* on the invitation to their Farmers Day at Elsenburg, which was attended by no less than 150 growers. They also used this to publicise their stand at Flora 93, which was held at the Good Hope Centre in Cape Town in September that year. It was later used for the 1998 IPA Conference in Cape Town. By the end of 1993, VOPRI published a list of 41 cultivars available from Elsenburg, with a further 8 cultivars suitable for landscaping purposes.

In 1994, the Botanical Society reported in Veld and Flora that they were allowed to have a stall at both Chelsea and Hampton Court Flower shows. Although the public were mainly interested in fresh flowers, they bought quite a lot of dried flowers. Interestingly, dried flowers never really took off in England in spite of interest shown at such shows, although for a short period of time they were used in the cottage industry to create wreaths and baskets for home decor.

More and more farmers opted for establishing proteaceae in plantations, particular as new cultivar material became available. New names appeared such as Spider, Aristocrat, Succession, Candles — it became a never-ending stream. At last the industry was starting to move in the right direction. A number of farmers added ferns to their repertoire as a popular item of 'greens'. The protea register had their hands full to record all the new cultivars and clones. But still, reports were published about veld harvesting, for instance the unique cable way system Andre Smith, a farmer near Ceres, deployed to get his flowers from almost inaccessible mountain slopes.

The Vegetable and Ornamental Plant Research Institute (VOPRI) changed its name to Agricultural Research Council (ARC). The industry continued to provide funding for research. This was collected via a levy on cartons. The system worked extremely well for many years. Calculations showed that about 80-90 per cent of the levy was forthcoming when compared to the export statistics. These statistics based on kilograms exported was provided by the Governmental inspection services and the two could be more or less correlated.

The ARC started a consultancy service, and appointed Gerhard Malan as consultant to the Industry. Nissan South Africa sponsored a four wheel drive vehicle, while BP donated fuel. A local garage sponsored the service contract. Real progress was now happening.

Willem Verhoogt of Bergflora addressed the AGM in 1992 and warned growers and exporters alike: *'It is clear to me that the competition on the international flower market is intensifying due to the fact that more third world countries with European capital and subsidies are dumping flowers on overseas markets in order to earn foreign exchange. The importer has a wide choice of supplies and suppliers and due to his decreasing profit margin is looking for supplies which are of outstanding quality. We will have to show that we have such quality available and that we are prepared to deliver a first-world service instead of behaving like a third-world country with all the consequences.'* It has never been an easy business and never will be, but the professionals will survive because they are able and willing to meet the challenges of the day.

Tony Rebelo had runaway success with the Protea Atlas project. He had mobilised enthusiastic amateurs, mountaineers, and botanists to go on regular hikes, field trips and climbs to identify, count, and report on proteaceae, first in the Western Cape and later further afield as far as KwaZulu-Natal. This mammoth task finally led to exciting new discoveries of plants believed to have gone extinct as well as numerous new species, which had never been described, in addition to increasing the range of a number species. The whole project culminated in the 'Protea Atlas'; a book no-one who has natural Fynbos should be without.

On the lighter side, a Cape Town florist was arrested in 1993 for attempting to export marijuana hidden amongst a shipment of proteas destined for London. What a reason for flowers to hit the headlines!

In 1993, Integrated Pest Management was a 'new' upcoming subject and was the theme for the Open Day in September. Recommendations were published in Roodeplaat Bulletin No. 38.

In 1994, the industry finally received recognition for Fynbos farming as a legitimate form of Agriculture. This was a result of discussion between the Department of Agriculture (Dr. Frans v.d. Merwe, Director-General of Agriculture), the Western Cape Agricultural Union, the Agricultural Research Council, and the Association. *'The commercial utilisation of Fynbos on an individual farm, whether privately owned or leased, shall be recognised as a farming enterprise. However, this recognition will exclude the exploitation of Fynbos on any land belonging to or falling under the jurisdiction of Nature Conservation and/or Forestry'*. I am quite proud of having achieved this on behalf of and for the industry.

*1994 Executive Committee L to R: Back row: Willem Verhoogt,
Kobus Steenkamp Colin Verwoerd, Cobus Coetzee, Barrie Gibson, Denis Shaw
Front row: Neels Beneke, Walter Middelmann, Maryke Middelmann*

It was a good year, as the industry agreed on a levy on dried flowers to support vital research and new projects.

There was, unfortunately, according to Prof. Gerard Jacobs, a lack of a joint marketing vision for the industry, with little involvement of the growers as to where and how best to market Fynbos. This message was repeated by Lampie Fick, MEC for agriculture at the 1994 SAPPEX AGM. Dr. Cobus Coetzee of the ARC also pointed out to South African growers that they would do well to emulate their Australian counterparts who plant proteaceae on a large scale for export to Japan who are known for their strict phytosanitary measures, while South Africans complained about rejections of insect-infested flowers picked in the wild.

International Developments

Portugal

In the *Sunday Telegraph* of 2 December 1990, Bob Pearson wrote an article about flower development and the difficulties Portuguese flower growers faced with hot summers and variable quality of alkaline soils. It is thought that Ms Flor Georges-Picot, who ran a landscaping business and who started experimenting with proteas, was the pioneer in the Algave. She had to take great care to know which varieties would grow under those conditions. By then of course, proteas were already being grown on Madeira.

Hawaii

The protea growers asked a local composer, Keith Haugen to write a pop song to promote their proteas. Sung by a cabaret artist from Sydney who was working in Hawaii at the time, the song received good airplay on the local pop stations. The composer sent copies of the song to then the President of South Africa, FW de Klerk, as well as to Nelson Mandela. I wonder if anyone still has a copy of this song.

Protea pua nani o Maui, Flower of Heaven
Protea, ever changing beauty
Your name for a god it was given
Protea, protea, protea, I want to touch you.

IPA

In 1991, Perth hosted the IPA Conference under the capable leadership of Maggie Edmonds. A large pre-launch party was held at Kings Park and Botanic Gardens the year before to draw attention to the event and garner support from the authorities. The event itself was held at the Parmelia Hilton with pre-conference tour northwards to Geraldton, while the post conference tour turned south to Busselton and Albany after the four-day conference. Walter Middelmann, representing South Africa again as member of the Board, reported on the status of the South African Industry and wrote a full report for SAPPEX news. His introduction was a good summary of the industry at the time:

We are still in a situation that, due to the nature of the protea industry in South Africa, statistics in the way Conference management has asked for are not available. While plantations in orchard fashion, as they of necessity evolved in other parts of the world, have also increased here, there is still a considerable part of the material sold coming out of natural stands, particularly in the form of 'greens'. Plants in a more or less mass-sown way also play a role. However, with greater availability and interest in growing some of the fine cultivars and hybrids developed here and elsewhere, growers become more specialised and use more discrimination.

It was to be his last year of service to the IPA although he continued to attend IPA Conferences thereafter until the year 2000 in Tenerife, at age 90. He introduced me to the Board, who accepted me as replacement, and virtually immediately I got the job of leading an investigation into the possibility of International Protea Standards, with instructions to report back in Harare.

Phil Parvin, Ruth Middelmann and Peter Mathews during the post conference tour following the IPA Perth conference, 1991.

After the successful IPA protea brochure, jointly worked on by Australia and South Africa, the IPA also organised for a wall chart to be printed in the style of the very useful Dutch flower wall chart. This was coordinated by Walter Middelmann, with major input from Willem Verhoogt (South Africa) and Brian Harris (Australia) on their respective countries' proteaceae genera, while Mr. Sugiyama in Tokyo handled the Japanese translation. Explanations and text were in English, Dutch, German, and Japanese. An early version had already appeared on behalf of the IPA, done by Proteaflora, showing both South African and Australian top selling proteaceae.

Dr. Phil Parvin and Dame Joyce Daws (IPA Chairman) launching the new IPA wall chart at the Perth Conference

IPA Journal carried the news that the internationally renowned author and gardening specialist Sima Eliovsen of South Africa, Honourable Member of the IPA died on 4 June 1990. She was buried on World Environmental Day.

Dave Tranter handed over the reins of the *IPA Journal* to Phil Parvin of Hawaii. Dame Joyce Daws and David Matthews both indicated to members of the IPA Board that they would stand down as Chairman and Secretary respectively after the next conference due to be held in Zimbabwe. They were keen to move the secretariat to South Africa, but to find somebody to take this on would not be easy! Lois Turnbull of Australia took over the responsibility of the journal from December 1993, when Phil Parvin was due to retire from the Kula Research Station in Maui and move to Florida. On Phil's retirement they honoured him by naming the cultivar garden at the Kula Research Station after him.

Then David Seaman of Zimbabwe announced that the next conference would be held in October 1993, which is normally a very busy export month for South Africans who begged him to change this date. A most amusing 'open letter' was written for SAPPEX News by Erich Schmollgruber in which he posed the question : **'Why would** *the Chairman of a committee formed to organise a conference want to exclude the very people from that conference for whom it was organised in the first place?'*. Mr. Seaman replied that seeing that attendance by South Africans at past IPA conferences had been so poor, he could not consider objections from South Africa regarding the planned date of the Conference, which was approved by the IPA Board in Perth. As a result there were a bunch of unhappy people who had already been precluded from attending two previous conferences: the conference in Australia where South Africans were not welcome because at that time sanctions were in force, and the one before that in California for reasons of cost because of the highly unfavourable exchange rate between the US$ and the S.A. Rand at the time. Here, they thought, was finally an IPA Conference they could attend, and it was at the busiest possible time! In the end there were only nine growers and seven scientists from South Africa who attended, which was a great shame.

I do not remember the reason, but Gert Brits and Cobus Coetzee were at each other during one of the Board Meetings, contradicting each other regarding ARC policy at every turn. Things got out of hand until I gave them a stern talking to and told them to stop embarrassing the rest of us. The Board itself was grappling with the issue of succession of Chairman Dame Joyce Daws and Secretary David Mathews. Various people were approached, but nothing concrete resulted and Phil Parvin was asked to do an around-the-world trip on behalf of the IPA to try and find a future direction for the organisation. This trip never happened, so the status quo remained until the next conference in Israel. Jack Harre felt that the problems of the IPA lay in the fact that the organisation had been highjacked by scientists, but on counting numbers, it was obvious that the IPA membership was pretty equally divided between scientists and growers — growers were just not as forthcoming to talk at conferences as the scientists! Eventually, it became clear that for the IPA to go on it would be necessary to bring it to South Africa, the home of the Protea. It was necessary to consult with SAPPEX before a final decision could be made.

The highlight of the Harare conference was the talk by Dr. John Rourke of Kirstenbosch on *'The conservation of genetic resources in the Southern African Proteaceae'* This talk became well known for the quote he used by Sir Otto Frankel: *'Cherish variation, for without it life will perish.'*

John's own synopsis of this valuable paper reads: '*The future growth and survival of the protea industry will depend on the continued development of new cultivars to meet the demands of increasingly competitive markets. Many of the most important new cultivars have been bred using genetic material from rare variant local races within widespread species that have fortuitously survived in nature or in cultivation. The maintenance of a sufficiently wide spectrum of genetic resources in the commercially important species of Southern African Proteaceae both in nature and in cultivated Genebanks, is essential to ensure the sustained production of new cultivars. This paper highlights ways in which the IPA can support the conservation of these vital genetic resources*'.

(Refer to Appendix J: The Conservation of Genetic Resources, JP Rourke)

Two years later and after lots of discussion, this talk resulted in the establishment of the Fynbos Genebank Trust to support expansion of the Genebank at Elsenburg. Hawaii also worked hard to garner support for the establishment of a *Leucospermum* Genebank on Maui at the research station there, but politics got into the way of that one.

Another important talk that was republished in IPA Journal was that of Alon Malter of the Ministry of Agriculture in Israel, whose talk was entitled: '*The European Market for Protea: Recent trends and lessons for further product and market development*', in which he emphasised the need for a market-orientated development program to increase potential and help to guide the industry in its future expansion.

On the work front, some of the IPA sub-committees who were charged to finalise projects, reported back. One of them was my own project on international standards. After a long hard slog, I deemed it virtually impossible to set or enforce international standards. Standards would be dictated by the marketplace and international treaties.

At the same conference, Robert Middelmann gave a short talk on the dried flower industry in South Africa. It was the first time the topic of dried flowers was on the IPA program, and it made everyone aware that dead flowers and dried flowers are two very different things. When this talk was reviewed in 2010, it was interesting to see that most dried flowers were then sold in bulk, whereas in 2010, a major portion was sold in small special packs. Drying treatments and handling have also evolved quite a bit since then.

Joan Sadie of the Department of Agriculture, South Africa, presented the preliminary first edition of the 'International Protea Register' with regard to the South African genera of Proteaceae. As an example I give below the entry for that perennial favourite and internationally popular Leucadendron as follows:

> Safari Sunset,
> Ld. Salignum x Ld Laureolum;
> origin 1962 New Zealand,
> Breeder/Owner J. Stevens/I.Bell
> Publications: Matthews & C. '83, Riverlea '86,
> IPA brochure '87
> Remarks: Introduced 1964. 1st International successful Protea cultivar

(Refer to Appendix K: Safari Sunset)

The IPA Board debated my proposal of direct representation from each association, rather than from individuals and to work towards affiliating associations worldwide through a membership fee based on the number of members per the association affiliating. This would promote better communication internationally, and give the Board representation of a larger group of people instead of only individuals. The Board agreed and eventually all associations around the world affiliated and delegated a representative Board.

In preparation for South Africa to take over the administration of the IPA, we approached the Reserve Bank for permission to keep an account in US$ in South Africa. In spite of offering every good reason, and with the support of a local bank, the Reserve Bank refused to make an exception (it might have been allowed if the figures ran into millions instead of thousands) and eventually David Mathews agreed to keep a US$ account in Australia for drawing on if necessary, while we would run a Rand account in South Africa for minor expenditure.

Israel

Meanwhile in Israel the Protea Working Group of the International Society for Horticultural Science arranged for a workshop at Neve Ilan in conjunction with the International Congress in Florence, Italy. The organizer was Dr. Jaacov Ben-Jaacov. A number of South African scientists were involved in the program. At the time the proteas, which were introduced to Israel in 1974, were limited to Banksias, Grevilleas, *Leucospermum patersonii*, *Protea obtusifolia*, *Leucadendron* 'safari sunset', and *L. Discolour*.

Zimbabwe

The Zimbabwe Protea Association had been formed in 1983 with growers just about everywhere, except the Midlands and Matabeleland. Because proteas had to be propagated from seed or cuttings, the quality was of course superior to those of the still predominantly veld-harvested flowers from South Africa.

Cecily Saywood of Glenellen farm was one of two people who attended Chelsea in 1990 on behalf of the Zimbabwe Protea Association. They exhibited only proteas that year, but thereafter were to showcase also other flowers grown in Zimbabwe for export.

Cross-border visits by industry leaders were infrequent, but very interesting to those who did. Newsletters were exchanged between the associations, and Renate Parsley of Parsley's Cape Seeds was well known to growers in Zimbabwe. After a round-the-world trip to visit her seed customers and to learn more about proteas around the world she wrote a brief but interesting report for publication in the ZPA newsletter entitled 'Protea: Where do you grow?'

A lot of publicity was given to the Harare IPA Conference after the event, which undoubtedly resulted in renewed enthusiasm to increase cultivation.

Australia

A very interesting exercise on estimated costing for protea plantation was published in 'The Australian Protea Grower' in 1991. In it, the author calculated that on 2000 plants he would make a loss of 17c per bush, while if he had 4000 plants, he would make 38 cents per bush. In the calculation provision was made for only one labourer, working two days a week to do spraying, pruning, weeding etc. Presumably the farmer himself did the picking, cleaning and packing. Hardworking bunch of people those Aussies!

In 1990, the Commercial Protea Growers of W.A. nominated James and Elizabeth Wood for the 'Horticulture 2001 Award,' which they won against stiff competition from other large agricultural companies. They certainly made an impact on the protea scene in Australia. By 1991, the Australian Protea Industry had drafted their first standards for cut flower, based on a draft done in 1986 by James and Elizabeth on which they drew from their experience over many years as growers for export in South Africa. The draft had been done in collaboration with the W.A. Department of Agriculture and the Department of Primary Industry. Changes were made to bring it in line with current international requirements of the EU standards for Horticultural Produce.

At the time, there were two Protea Associations in Western Australia, which amalgamated in 1991 as the Proteaceae Producers Association of W.A. Then a Steering Committee drafted a document for a proposed Flower Industry Association of Australia. Well, this did not go down to well, as there was nothing in the proposal regarding estimates of the cost of running such a body and its subcommittees. In its response, the PPAWA stated that they wanted nothing to do with such a proposal until proper consultative measured had been carried out and met with PPAWA approval. Elizabeth Wood wrote: *'It is worth considering that successful growers almost always design their own business strategy. To be successful it is fundamental to take charge, to be responsible and not rely on others or a 'board' for motivation and discipline'*.

U.S.A.

In 1992, the California Protea Association brought out a useful Beginners Handbook entitled 'So you want to grow Proteas!' Although of course the figures quoted regarding acreage planted, cost of establishing etc., is long outdated by now, the basics are still valid, and is a good guide for new entrants.

An interesting newspaper cutting reached us from Pauma Valley, California, where in late February the San Luis Rey River was so flood-swollen that a helicopter service was called in to move Ben Gill's proteas from Silver Mink Farm out to their market. The helicopter moved 42 x 50 lb boxes of proteas across the river to a waiting truck. Some of the other farmers followed suit and successfully got their produce to the market in time.

A friend of the Middelmann family, Dr. Cecil Eschelmann from California, sent us the pages from *Sunset Magazine*, depicting a large wreath of fresh proteas and banksias under 'Happy Holiday Ideas' featured on the front page and various smaller wreaths and bouquets on the inside pages.

Chile

Visitors to South Africa frequently bought a few packets of seed to try at home. One such visitor was a Lynn Rodriguez who visited South Africa in 1994. A flower farmer in La Liqua, she had success in propagating these and started a small plot of proteas. She eventually corresponded with South Africa as she was keen to try not only items of their indigenous flora for the international market, but also some of the proteas for fun. How she got hold of my name I have no idea, even less how she decided that I would be the one that could supply her with plans of drying sheds. She wrote that Dr. Ben Jaacov of Israel had been contracted by the Chile University to study if it was feasible to propagate proteas in Chile, and he had visited her as he was interested in seeing how she had successfully grown proteas. Of course! That is how she got my name!

Industry Developments 1995 - 1996
South Africa

In 1995, the Department of Nature Conservation invited Fynbos farmers to a seminar at Grootvadersbosch near Heidelberg (Cape). Conservation/Botany scientist Prof. William Bond warned that even harvesting 50 per cent of a plant could be harmful in the long run. Harvesting levels should be based on how many seedlings come up after a burn and furthermore one should consider desired plant density required before make a decision on harvesting levels. Suddenly the interest in the effect of harvesting on the environment was receiving attention of conservators and botanists alike. Leading this field were Jan Vlok of Cape Nature Conservation in De Rust near Oudtshoorn, William Bond, and others like Penny Mustart of the Institute for Plant Conservation, based at UCT. Pat Holmes came on the scene, focusing on rehabilitation after clearing, as well as conserving of seed stores and rehabilitation and re-seeding after major earthworks. They continued to publish interesting facts like insect and plant interactions. There was also a warning that disturbance of veld would lead to invasive aliens gaining a foothold.

Phil Parvin was making arrangements to come to the Cape, together with Jim Heid representing Hawaii Growers, with a view to foster better cooperation with the South African Industry on cultivar exchanges via the ARC's Fynbos Unit, led by Cobus Coetzee. It was also the intention to pay a courtesy call to Maryke at Botrivier! After their visit we received their thank you, Phil's customary pink flamingo and a 'gater' from Ron. They obviously love their Florida home!

The Fynbos Forum under Chairmanship of Christo Marais ran a workshop at Stellenbosch. A large number of monitoring programs and surveys had been done by individual institutions, but there was no integration between institutions to prevent duplication of monitoring programs. It was decided to compile an inventory to summarise all data that could be used in monitoring, management programs, and other activities in the Fynbos Biome. With support from SAPPEX and IPA it was possible to publish Volume I in 1995 and subsequently Volume II also saw the light.

Leading researchers and conservationists, during brainstorming sessions at Fynbos Forum, decided that help was needed to protect the Fynbos from Alien Invasive Plants. Dr. Brian van Wilgen of Forestek together with the Institute of Plant Conservation put together a presentation to bring across this message: *Cape Town's water supplies are seriously threatened by alien plants (introduced from America, Europe and Australia) which invade catchment areas and use much more water than the indigenous Fynbos. The threat, which may result in losses of up to 30 per cent of Cape Town's water supply, is aggravated by the fact that funding for the clearing of alien plants has declined substantially over the years.*

High ranking government and provincial officials, Fynbos Forum representatives and key policy makers were invited to attend the presentation on Friday 2 June 1995 in the Auditorium of the Cape Provincial Administration Building in Cape Town. Straight after the presentation, Prof. Kadel Asmal, then Minister of Water Affairs and Forestry, stood up and declared that funds would be allocated to eradicate alien plants on a large scale. The eradication would serve three main purposes:

- Delivery of water: Immigration to the Western Cape is increasing, putting huge demands on limited water resources.
- Protection of biodiversity: The uniqueness of the Flora Capensis with 8574 species of which 86.2 percent are endemic, is recognised as the world's 'hottest' hot-spot.
- Job creation: Training teams to eradicate alien invasive plants would lead to the creation of over 800 jobs for ten years. This investment would ensure that the Cape is better able to meet its water needs.

Nature Conservation's Christo Marais was charged with getting the project off the ground. Christo proved that his capacity for work is nothing short of amazing. He made a great success of this long-term project and within a relatively short time, the yellow T-shirted 'Working for Water' teams became a common sight in the Fynbos. Training was not only in recognition of plants and their eradication, but mountaineering skills, use and maintenance of tools, transport and support services such as crèches for working mothers, were all part and parcel of the project. So successful was this project that it was copied in other provinces that also had their unique invasive plant problems. In this way a large number of unemployed people learnt valuable skills and earned a stable income. The program was so successful that it received attention by overseas scientists who used lessons learnt from this program for their unique invasive plant problems. Christo subsequently expanded the program to include 'Working on Fire', and 'Working on Woodlands'. In between planning and implementing the program, he even managed to obtain his PhD. He received the Cape Action for People and the Environment Gold Medal in recognition for his work.

The 'new' South Africa was working hard at integrating the 'lost generations' into society. To this end the ARC embarked on small farmer development in the Fynbos. Numerous pickers and interested people from the old Mission stations at Genadendal, Elim, and elsewhere were given training in Fynbos propagation and plantation management. Plant material was supplied free of charge or financed via government structures. Sadly, in spite of one farmer winning the 'small farmer of the year' award, this project, as well as the others, eventually petered out.

1995 was quite a busy year, as it marked the launch of the Fynbos Genebank Trust — a joint ARC/SAPPEX initiative with support from IPA. The launch was widely reported, the one I particularly enjoyed was in Chronica Horticulturae, which erroneously gave me the title *Dr.* Maryke Middelmann! Although it was hard work to get it off the ground, the launch on 25 April was really a prestigious function at the magnificent recently restored Elsenburg Manor House. Particularly Emmy Reinten of ARC was extremely helpful and gave more than could be expected to make it such a wonderful event. I will never forget entering the imposing building to be greeted by a table decked in black velvet, with glowing glasses of sherry, and a single perfect *Protea holiserisia*, one of the most vulnerable of protea species. The event was attended by Phil Parvin and Jim Heid of Hawaii, Dr. Alan Heydorn, Brian Huntley and John Rourke of Kirstenbosch, and a host of other VIPs.

At the launch of the Fynbos Genebank Trust: Prof Phil Parvin, Maryke Middelmann (Chairman SAPPEX) and Dr. Cobus Coetzee (ARC Fynbos) (Agricultural news 5 June 1995)

Dr. Alan Heydorn opened the proceedings with a short talk on the historical connection with WWF through the legacy of Protea Heights. Phil Parvin stressed the importance of having a large pool of genetic material and that it is of international concern that the Fynbos Genebank is preserved and supported. Gail Littlejohn explained the function of the Genebank and I made the appeal for support. In my short talk I reminded the audience of John Rourke's wonderful talk in Harare and his quote by Sir Otto Frankel '*Cherish variation, for without it life will perish*'. The importance of maintaining our genetic diversity, not just for future generations, but for improving a potentially lucrative crop was the reason behind the Genebank. The aim was to generate sufficient funding for the interest on the Trust to be enough to cover the annual research and maintenance bill of the Genebank. Trustees were invited. These consisted of the ARC, SAPPEX, as well as an IPA representative and one 'outsider'. We considered ourselves fortunate that John Rourke agreed to fulfil this role.

The draft guiding principles for the Fynbos Industry were published for members' input as required by the Department of Agriculture, as a condition for recognition of the industry as an agriculture pursuit.

At last, also in that year (1995) the U.S.A. decided to lift sanctions against South African proteas, under certain conditions, the main one being that there would be a considerable improvement in the rejection percentage.

The provincial standing committees for Agriculture and Environmental Affairs were hosted by SAPPEX with the objective to show them how Fynbos provides jobs in the Western Cape and how this can go hand in hand with sustainable utilization. It did much to enhance the profile of both the industry and SAPPEX.

One of the main concerns was still that of runaway fires and again in 1995 a number of farmers were hard-hit. A holistic fire-management process within the Province was a high priority, particularly involving Cape Nature Conservation areas that were not taking the correct fire protection measures due to lack of funding. A coordination committee was later formed on which SAPPEX had representation. This committee talked a lot, but didn't achieve much, so I resigned! Since then Fire Protection Committees have come into being within certain physical boundaries incorporating farming and conservation areas. Still later 'Working on Fire' teams saw the light, but fires still cause devastation and mayhem.

Gerhard Malan (Landbounuus 1990)

Gerhard Malan published a full report on the Fynbos Ornamental Industry (1993-1995), sponsored by SAPPEX. The 'Malan Report' as we normally referred to it, is the only in-depth study of the industry I know of.

In 1996, a hawker was arrested for illegal trade in proteas. He was acting for a restaurant owner, who had a Fynbos smallholding in the Overberg. It was once again one of those cases where the person arrested was innocent, while the owner was one of so many who thought that it was a waste of time to belong to an organisation like SAPPEX. If he had joined for the paltry membership fees, he would have known about the necessity for permits.

An unexpected and exciting opportunity came my way in September. We were at the Josephine Mill near Walter's flat where we had a gathering of Ruth's friends, family, and business associates from various walks of life to celebrate her life. After she died on 15 June 1996, Walter did not have the courage and stamina to organise such an event, and when I suggested that we should give all the people she knew so well an opportunity to say a proper farewell and what better date than her birthday on 28 August, he agreed. He put together a display of items including representative frogs from her huge collection, her music, her Botanical Society show certificates, pictures over her lifetime, historical documents, and of course her life's work on proteas in its various forms.

It was at that gathering on Saturday that Colin Verwoerd of Magnifica CC approached me and asked if I would be willing to go to Moscow. He had been contacted by the South African Embassy there, who was keen to have a presence of Fynbos at the Flower Show in Moscow. Well, I told him I would be more than willing and asked 'when?', and he said: 'you will have to leave next Wednesday'! My response was, 'you organise me a visa and I'll be there'. By Monday morning the Embassy in Moscow phoned to ask me to fax my passport details and to deliver my passport to my travel agent who would send it by courier to the Embassy in Johannesburg. Then there was a mad scramble to get flowers together, because we were doing dried flowers from Honingklip and fresh flowers for Colin Verwoerd. It was not a SAPPEX stand, but a commercial stand. Therefore, Colin had to write a story about his business while I wrote a story about Honingklip, both were ably translated and duplicated by the Embassy in Moscow.

By Wednesday morning my passport and ticket were ready at the travel agent and off I went by KLM to Schiphol and from there to Moscow. I was picked up by embassy personnel, whisked through the diplomatic channel and taken to view a number of hotels. I made my choice and settled in for the night. It was a hotel for local tourists and none of the staff could speak English, but that did not deter me. It was in close proximity to the All-Russia show grounds and that was important for me. The embassy undertook to clear and deliver the flowers to the stand the next day, so that I could do the necessary. Not only did they deliver the flowers, but I also sent them to buy buckets and vases so I could make up some display, and they helped with putting up posters. I was the envy of the stands around me and I really bragged that 'this is what our government does!' I had hired the services of a translator, a pretty young little lady, and it hardly cost anything.

I have to mention that I was also the envy of the other stand holders because of the huge interest in the South African stand. So many local botanists touched the flowers in wonder. They had not thought they would ever see a real protea, leave alone touch it. I had that same feeling of wonder when I first set foot on the famous Red Square outside the Kremlin.

At the end of the show, my translator insisted that the flowers should be sold. That's not something I'm good at so I told her to go ahead and she should see what she could do. In the end the proceeds paid for her services, a night out to dinner for both of us, a camera for her, as well as a cash bonus. I even had a stack of Rubels to take home, which I changed for US$ at the show grounds. The large blue/white vase I received from the judges at the prize giving, I gave to the S.A. Embassy in Moscow as thanks for their wonderful assistance. Years later, Witz du Plessis who had been so very helpful, came back to South Africa to head up the Department of Trade and Industry office in Cape Town.

This story would not be complete without a bit about the hotel and its food. After a few days of wet towels I firmly believed that my towel was being hand washed and left on the rail to dry. Then I noticed that the rail, which doubled up as a towel heater, was leaking and that had made my towel so wet. There was an interesting tap which you swung left if you wanted a bath and right if you wanted water in the basin — nifty that! I heard later that I was lucky to have a plug for the bath. In the dining room things were not much better. The first day the food was plentiful: bread, cheese, eggs, yoghurt, and a variety of dishes for dinner. By the last day, all that was left was potato salad for both breakfast and dinner! The menu of course was only in Russian, and when I asked if anyone could speak English or German, I met a gentleman from Gotha in the former East Germany. Getting chatting (me in my best German), it turned out that he was looking for a source of dried flowers and there I was, so we had lots to chat about. Having made a friend, he frequently brought me fresh coffee that he brewed at his stand. Robert and I subsequently visited them at their place, and they came to South Africa and stayed with us for a few days. In 2010, we still sold to a customer in Moscow, from a contact made during that trade show.

Apart from that highlight, SAPPEX participated at an exhibit in Cape Town called Florex 96 where Hans Hettash and Marius de Kock of ARC were of great help. Then SAPPEX went up north to support the ARC stand at Gardenex in Pretoria.

Also in that year, both SAPPEX and ARC planted a tree, with much pomp and circumstance, at 'Die Bos', the entertainment area in the Elsenburg property, on the occasion of National Arbour Day. I wonder if 'my' tree is still there.

The new Cape Flora brochure was published by SAPPEX, this time it was compiled by ARC with close cooperation from SAPPEX and its exporters. One of the results of closer cooperation with the ARC led to ARC news being included in the SAPPEX Journal. And, big event, SAPPEX announced that they would host the 1998 IPA Conference in Cape Town.

International Developments

IPA

In 1995, the first draft guidelines for Conference Organising were drawn up, as were Guidelines for Directors. Discussions were held between David Mathews, Cobus Coetzee, Gail Littlejohn, and Maryke Middelmann on the future of IPA and key questions were identified for a membership workshop.

Protea News, a separate publication for the International Protea Working Group (IPWG), was published twice a year from August 1984 until 1991 when it was stopped for lack of funding from the Department of Agriculture's Vegetable and Ornamental Plants Research Institute (VOPRI). In 1995, it was resuscitated and was for the first time published within the IPA Journal. In that particular issue they put out details regarding the Israel Conference, with a post-conference tour within Israel to take in the most famous sights. They offered an option to extend this with a trip to the Netherlands flowers auctions. It was an opportunity to visit yet another country and see how things are done there.

IPA embarked on a scholarship program and in 1995 supported four research programs, one in South Africa, another one in Israel, and two in Australia. By early 1996, there was concern about the Conference that Israel was arranging for that year. Apart from the initial announcements in the IPA Journal the previous year, nothing had reached the associations or the IPA members. The problem seemed to lie in the fact that they had appointed conference organisers, and clearly for the IPA a hands-on approach by the host country's representative is a prerequisite for success. In spite of much cajoling from Walter, the conference looked like it was going nowhere. When Duby Wolfson stated in June of 1995 that so far they had only thrity-one overseas delegates and that they might have to make a difficult decision, the wires went hot backwards and forwards between Cape Town (Maryke and Walter) Australia (Joyce and David) and Israel (Duby, Boaz and Ben-Jaacov).

In the end, the conference in March 1996 was a success with participation from delegates from many countries. On the final evening, we were entertained to Israeli traditional dances and songs, which quite a few of us joined in, much to the surprise of the locals. Walter and Ruth Middelmann were made a special presentation as 'Parents of the Protea' by the Flower Board of Israel. Sadly it was Ruth's last conference, because in May that year she suffered a major stroke from which she did not recover.

Presiding at the IPA's Israel Conference 1966: Dame Joyce Daws (Chairman)

The post-conference was amazing for us first-time visitors to the Holy Land. I thought Hans Hettash and Gail Littlejohn were nuts to climb up the Masada instead of taking the cable car. I walked/jogged the downward journey and was stiff for days! And who will ever forget us smeared in Dead Sea mud! The yummy garlic infused olives at the Museum will always remind me of the Dead Sea scrolls we inspected there. The history of this place is of course staggering and one wonders what lies under your feet . . .

At the Israel Conference I was elected Chairman, and saw this as a suitable chance to ask, at the first Board meeting, (still under Joyce Daws) whether South Africa might take the next conference instead of Tenerife as scheduled. It had not been possible for South Africa to host the IPA due to our political situation, so we thought it was high time. We needed to put on strong conference in order to stimulate interest in the IPA and it would be easier for Tenerife to ride on the momentum, rather than to take over after two successful but relatively poorly attended conferences. Fortunately, the Board agreed wholeheartedly.

The first journal after the Israel conference included a questionnaire on the future of IPA, and a decision was made to incorporate IPWG news into the body of the journal instead of as a supplement. The results of the questionnaire were published in the next journal and many of the issues were debated via numerous memos between the Board.

Industry developments 1997 - 1999
South Africa

In *Nature* Volume 386 an important article was published, entitled 'The value of the world's ecosystem services and natural capital'. This article brought a new perspective on Fynbos: how nature provides 'services' without which life on earth would cease to function. Nature provides, for instance, climate regulation, water regulation, soil formation, nutrient cycling, genetic resources, opportunities for recreation, and a host of other 'services' that we just take for granted. Attempts to put a monetary value on this are of course very difficult, but here in South Africa like elsewhere, such calculations were made, which makes a convincing argument for conservation of the natural habitat. Fortunately, it has been possible to conserve large tracts of the environment through the type of scientist that I met through the Fynbos Forum. Richard Cowling in particular, to my mind, played a leading role in this.

The Agricultural Research Council was invited to Washington to present a series of lectures at the Arboretum as a result of the signing of a Science and Technology agreement between South Africa and the U.S.A. I was fortunate to be invited along, prior to jetting off to Holland for Hortifair, probably the largest floral trade show in Europe. The ARC had rented a large apartment in Washington where we camped out in style. (I was the lucky one to get the huge main bedroom with giant Jacuzzi to myself.) Staff from Roodeplaat and Elsenburg participated as a number of people up at Roodeplaat were working on indigenous bulbs, mainly Lachenalia, originating in the Fynbos. Elton Jepthas, who had received a US-Aid scholarship and had spent time at Belltsville and Denver, was with us the last few days and camped in the lounge. His heap of stuff in the corner made the place look like a squatter camp, so I suggested he keep his belongings in my very large walk-in-cupboard.

Unique Cape Flora leaflet designed for Washington and Hortifair

One of the highlights of that trip was that it coincided with Richard Cowling making a presentation to the World Bank on the Cape Floristic Kingdom to explain the necessity to conserve this Biome for future generations. Cobus Coetzee and I broke away from Arboretum activities to witness this presentation — Richard was surprised to see us! I might add that it turned out to be a very worthwhile visit for Richard and it was not long before substantial funding came into the Western Cape for conservation projects. Much later (2004) SAPPEX was also a beneficiary of funding for a small project on Sustainable Harvesting.

To get back to the Arboretum. 'Floral Gems: Blossoming Wealth of South Africa' was the feature theme. During the duration of the South African exhibit various lectures were given by ARC staff. We decorated the Arboretum entrance with fresh and dried flowers, artefacts, posters, and water colours of proteas. Very effective was a pedestal, covered in blue velvet, and a stunning crystal vase in which was placed a perfect Protea cynaroides; the 'flowering gem' thought up by Mariana Purnell, Agricultural Councillor of the South African Embassy. In the *Landbouweekblad* of June 1998 it was reported that all the fresh flowers that were sent to Washington had been meticulously inspected for insects, and those that were found were removed by hand, due to the strict phyto-sanitary measures enforced by the U.S. Department of Agriculture (USDA).

In between, we managed to pay a visit to the White House (a must when you get to Washington). We met up for a workshop with the ARC's American counterparts, also to discuss collaborative research. We were invited for a BBQ later that day to the home of the Floyd Horne, Secretary for Agriculture (equivalent to the South African Minister of Agriculture). There was also a cocktail party at the Arboretum, with South African snacks. For many, the highlight was a visit by Tipper Gore, wife of Al Gore, Vice-President of the U.S.A. and a real friend of South Africa. During her visit a new *Ornithogalum;* white with a green centre and a dense flower head, named in her honour was presented to Tipper by Gail Littlejohn, the breeder of this cultivar.

I found it a rare privilege to present Gail's final talk, the last one on the program on 29 October, as she had to make an early departure for New York. It gave me the opportunity to thank the Director of the Arboretum, Tom Elias and his staff and the public for their warm welcome.

If you think it was all fun and games, you should have seen us up and about early to go on foot, train, and bus, to get to the Arboretum on time, with the same trip back in the evening. Shopping for sustenance was also an expedition of note. Fortunately, from time to time some of us got a lift with the Tom who more than often got lost and took the incorrect turn-off. He spent quite a bit of time with us outside working hours and we suspected that this bachelor had his eye on one of the ARC girls!

Subsequently, the Fynbos Unit took up the name 'floral gems' for its superior cultivar releases, not only of proteaceae, but also Lachenalia, and Strelitzia varieties.

From Washington we flew straight into Holland where ARC/SAPPEX shared a stand at the South African National Pavilion in Aalsmeer. What an eye-opener such a flower show is. Firstly your head spins from the colours and scents, but to see first-hand the immense trouble and expense people go to in order to make an eye-catching stand really is mind boggling. It is also amazingly hard work and very tiring to stand there all day under artificial light with a continual stream of people and constant noise. So the infrequent breaks one can take when there are two people on the stand, make it far more bearable as well as giving one an opportunity to have a look at other stands and exhibits. What is particularly nice of course is to come across people from other countries that you have known in the trade for so long.

In 1997, Gail Littlejohn had obtained EC funding for research into crop production aspects of proteaceae to improve the product quality traded in the international ornamental community, with South Africa, Zimbabwe, Spain, and France as partners. Fifty-six per cent of the budget was allocated to South Africa. Many scientific publications resulted from this research that serve as a reference for future studies. But of course this also led to dissatisfaction amongst some short-sighted members who did not want to share research results with France, Spain, Portugal, Zimbabwe etc, as prescribed by this substantial funding. This same funding gave the ARC Fynbos a respite and stalled possible closure.

In 1998, the Department of Agricultural Economics, University of Stellenbosch, published its findings on trends and factors that affect the supply and demand for cargo space for Fynbos exports from South Africa (compiled by Nick Vink) in which it was made very clear that airfreight is and would remain the limiting factor. This bottleneck could partially be relieved by sales to other countries, but with the prospect of Europe remaining the main market, it was suggested that sea freight under controlled conditions should be investigated.

In 1999, I was invited by the senior lecturer, Mohammad Karaan, to give a lecture on the 'Economics of Protea Farming' to the Agricultural Economics Honours students at Stellenbosch. Here was I, totally untrained, considered an authority, quite flattering! Prof. Karaan subsequently rose to the position of Dean of the Faculty of Agri-Sciences at Stellenbosch and was later appointed to the National Planning Commission. Our paths continued to cross from time to time at lectures and seminars. The man has the most incredible memory and is much admired for speaking his mind.

In 1999, a Wild Flower Conference was held in Australia. I took the opportunity to attend and came back brimming with new ideas. I was allocated funding for IPA Chairman's travel up to 2001 of which I used a portion to pay for expenses in Melbourne, while I paid the ticket from my own funds. It was an ideal opportunity to learn how the Australians go about identifying new flowers and products and to find out more about product certification. It also gave me an opportunity to talk about cultivar development in South Africa and to give a short talk on the IPA. As an added bonus, I made contact with the Western Australia protea growers who were not part of the national body. There just is no substitute for personal connections, either locally or overseas.

On a completely different note, something quite unusual happened in 1997 — the sentencing of a Pretoria man for stealing books from the NBI's Mary Gunn Library in Pretoria. Almost two years passed between the date of the theft and the date on which sentence was passed. The Middelmann family was involved because of Walter Middelmann's passion for collecting Africana and Botanical books. It all started when Walter purchased a rare book from Mr. Daniel Lloyd, well-known and respected book dealer in London. A fellow Capetonian had noticed that this book, as well as six others which were listed as missing in publications of *Forum Botanicum* and *Bothalia*, was offered for sale in Mr. Lloyd's catalogue. The books were duly inspected at the premises of Mr. Lloyd and found to be the missing books, but three of the missing books had already been sold. One had been sold to Japan, one to somebody in London, and one to Walter Middelmann. Walter, on hearing this, phoned Brian Huntley who also thought that this must be one of the missing books. Brian in turn contacted the head office of the National Botanic Institute who confirmed that the copy Walter had bought was one of the books stolen from the Mary Gunn Library. This set in motion correspondence backwards and forward between Walter and Mr. Lloyd, and Walter and the National Botanic Institute. There was a police statement, and subsequently, a court case in Johannesburg where Walter, then already eighty-seven years old, had to appear in person. The perpetrator, aged 43, was found guilty and was suitably sentenced. Some of the books were returned to Pretoria, but others remain missing to this day.

Elwen van Schouen and Maryke went to the Foire de Nancy in France on behalf of SAPPEX. Elwen was fluent in French, which was a prerequisite for a successful promotion event. We thought we were just exhibiting, but learnt that it was actually a fair where sales took place. We sold, and plenty . . . We had no props, but scrounged up old packing crates, wooden pallets, and a table, bought a sheet, borrowed a few containers and baskets, and turned our stand into quite a rustic, eye-catching one. It was a killer: we were busy from the time the doors opened till the doors closed, but we succeeded in making proteas known. We had to ask for more flowers from our growers as we had a heck of a job keeping up with demand. The boys at home were really on the ball and organized to get the flowers to us in tip-top condition. It was striking to see the huge difference between superior Cape Greens grown in plantations compared to veld harvested species.

Fortuitously, good timing meant that I could make a detour to the Chelsea Show for which I had received a complimentary ticket from

Kirstenbosch. I was keen to go, also because I could take the opportunity to visit my son who was working in London at the time. What a venue and what a show. If possible every flower lover should go to the Chelsea Show at least once in their lifetime. I was also fortunate to have been invited to South Africa House for a cocktail party where our Ambassador at the time, Cheryl Carolis, held an auction of fresh protea bouquets for the high society of London, in aid of the Mandela Children's Fund. I could not resist buying a bunch for the hotel where my son had arranged accommodation for me.

On the SAPPEX front the Executive Committee felt that we should give the members better value for money, and we embarked on a series of full colour leaflets on diseases, insects, cultivars, genetics etc. It made for a wonderfully useful series for growers. SAPPEX registered its name and Logo as a Trade Mark. SAPPEX also joined the electronic age and got its first, if somewhat tenuous link to email. Not only that, but the first tentative moves were made to create a web site. Lucky for SAPPEX, one of our members, Charles Oertel, offered his services to work on this project.

For some time SAPPEX had been running a section of ARC-Fynbos news within its journal and in 1997 published a short resume with a picture of all the researchers and their functions, which the members found very useful.

In that year a further analysis of the industry was published on information gathered by Gerhard Malan who had by then resigned from the ARC and went on his own as a consultant to the industry.

The ARC, after considerable discussion with SAPPEX and its members, finally allowed regional testing of cultivars, which led to various nurseries being appointed as bulking up nurseries, because the ARC nursery could not cope with the demand and was not seen to be equitable in its distribution of cultivars.

1998 saw the new deep red *Protea cynaroides* cv Madiba on the front cover of the *Landbou Weekblad*, with a report of new cultivar releases.

Gert Brits, the well-known and highly respected researcher at ARC, who had started the Genebank activities and cultivar development first at Riviersonderend and later at Elsenburg, took an early retirement due to heavy budget cuts at the organisation, which resulted in the number of posts being reduced. Gert was known throughout the protea world and he was given recognition at the Cape Town IPA Conference for the role he played in the development of proteas as a horticultural crop. He continues to develop proteas as pot plants for a commercial venture near Stellenbosch.

In 1998, Dr. Marie Vogts, the grand dame of protea cut-flower research, who had laid the foundation for the modern protea industry, died at age 90.

A milestone in the history of SAPPEX was reached in that year, when SAPPEX published its hundredth edition of the newsletter/journal. SAPPEX also embarked on a promotion program that would prove successful beyond all expectations. The exporters were keen to see if they could increase sales outside the normal marketing window by popularising the use of proteas. Tiel Bluhm, past CEO of Florimex, was appointed to take on the task of promoting the use of proteaceae with normal cut flowers and he did a fantastic job. Over time, Tiel succeeded in getting the German Florist schools on his side. He made contact with them during the IPM shows and arranged for a series of shipments to florist school competitions to highlight how to work with proteaceae.

For Frans and Meg Gerber of Forest Ferns, 1999 was a special year. They had decided to open their enterprise to the public and built a very informative Fernery and tea-room for Eco-tourism. On a tour of the Eastern Cape, President Mbeki visited them to familiarise himself with opportunities in the Eastern Cape. From what Meg told me later, the security checks were very thorough. Mr. Mbeki was an easy visitor, quite laid back and dressed casually for this event.

The ARC and SAPPEX again exhibited at the fourth Gardenex and Growtech exhibition in Randburg, where the stand won a gold medal.

The ARC tabled a four-year research program so that members could understand the long-term projects and the necessity for funding. SAPPEX started to investigate Quality Management Systems like ISO, the Flower Label Program, and Milieu Project Sierteeld. Certification of quality had started in the fresh flower trade and we realised that the time had come to assist growers in their decision in which program to participate.

Flower Valley came on the scene in 1999. Land was purchased with money provided by Flora and Fauna International and other UK sponsors, who made it possible for Flower Valley to work under ideal circumstances without having to make a profit. They entered into partnership with neighbouring landowners and soon made a name for themselves in conservation circle as having 'piloted and replicating models that integrate conservation and community benefits'. They aimed to record the effect of harvesting in a scientific manner and convinced a number of conservationists that only they know how to harvest sustainably. Unfortunately, in the process they alienated many farmers who had practised sustainable harvesting over many years. A number of them feel that it's all good and well to 'do-good' when you seem to be able to generate sources of funding on good public relations work. It is harder when you have to make a living! In the long term it is possible that their proposals for sustainable harvesting, although unproved outside their immediate area, can cause trade to be even more difficult than it is today.

International Developments

International Protea Association

In July 1997, the sad news reached us that Peter Mathews, the driving force behind the formation of the IPA, had died on 29 June. He had kept his interest and involvement in the IPA to the end. He was already not well in February when he wrote: 'I am still in the land of the living even though I am having a few health problems'. The following notice was placed in the Melbourne newspaper by Joyce Daws on behalf of IPA:

> **'Mathews, Peter**
>
> Founder Chairman of the International Protea Association.
> Man of vision, inspiration to us all.
> Farewell from friends and colleagues in the Association
> and the wider world of flowers.'

All the protea people knew Peter as a leader in the protea community worldwide. His family carries on under the Proteaflora name and has made a substantial donation to the IPA for research in his name.

In order to get support for the IPA to be represented on the Board by Associations, rather than by individuals, I made use of one of the business trips to the Far East and America (on a round-the-world ticket) to visit growers and associations in Hawaii, the Big Island and Maui, and California where I outlined my thoughts for the future of the IPA. In Japan, I made a point of meeting up with Mr. Sugiyama who had done so much to promote proteas in Japan, including a Japanese translation of the IPA Proteaceae poster. I was quite 'miffed' when the Hawaiians were more interested in asking Robert about dried flower production, than to hear about the IPA. It was a great visit and we met a lovely crowd of people who went out of their way to make us welcome.

In 1998, at the International Protea Conference and Protea Working Group Conference in Cape Town, 150 delegates attended from thirteen countries. It was my first term as Chairman. My first duty was to ratify the proposal for Board delegation, which was duly passed. The opening address by Richard Cowling on conservation of the Cape Flora: 'Strategies for the Future', gave everyone food for thought. A major coup was that funding was now available for research and conservation from the Global Environment Facility (GEF). (Remember that earlier report about Richard talking to the World Bank in Washington?)

In 1999, I was, in my capacity as Chairman of IPA, invited to Zimbabwe by ZPA to attend their AGM in Vumba. The AGM was held at the famous Leopard Rock hotel in one of the nicest locations in Zimbabwe. IPA business also took me to Tenerife to interact with the organisers of the year-2000 conference. Louis Turnbull, who had so successfully compiled the IPA journal resigned in July 1999, leaving it to myself and Margaret, the secretary, to add the duty of the journal to our other work.

New Zealand

One of the best known names around the world was probably that of Jack Harre who died in 1998, aged 67, after a long illness. He certainly was well known in IPA circles, also for his numerous publications on proteas.

In 1998, Gold Strike, a new variety of *Leucadendron* became the winner of the 'Most valued plant award for 1998' in New Zealand. They were sold there under Plant Variety Rights. New Zealand also developed *Leucadendron cv* 'Jester' with PVR by Duncan & Davies, who marketed this new variety through Torwoodlee Flowers of JS and ML Pringle.

U.S.A.

A protea festival was held at UCLA's Santa Cruz Arboretum in April 1998, over a period of three days. Speakers from Australia, South Africa, and Hawaii gave lectures as did a number of local speakers. The program also offered tours of four protea gardens, plant and book sales, and an art show.

The news of the death of quite a few well-known pioneers in protea cultivation was reaching us in quick succession. In 1998, we heard of the death of Ray Schatz, who was one of the original growers in Escondido where he established proteas around 1964. He shipped proteas all over the world from his fifty-acre ranch. He and his wife, Barbara undertook study tours in 1978 and 1981 and when he retired from growing in 1980, he offered consultancy for other growers.

Dean McHenry of the UCSC Arboretum died in April of the same year. Slowly but surely Walter Middelmann who knew all these pioneers from frequent overseas travels, was losing all his old friends.

Netherlands

The name of Klaas van Zijverden as General Director of the OZ Group in Holland is probably well known in floral circles worldwide. In 1999, Klaas was bestowed a Knighthood in the Order of Oranje Nassau by Queen Beatrix of the Netherlands, for his leadership in his community, including a care centre for the aged and services to the church. The same year OZ Group amalgamated with the Van Duijn Group to form the by now well-known Dutch Flower Group.

The firm of Oudendijk in Holland, one of the firms who pioneered imports of South African indigenous flora (Fynbos) in the 1960s, celebrated their fifth Anniversary. Both Hans and Niek Oudendijk are staunch supporters of the International Protea Association and have sponsored quite a number of the IPA's activities.

Japan

In Japan, publication of Protea publications with Japanese text continued, with credits to the Botanical Society, the IPA, Walter Middelmann of South Africa, and Brian Harris of Australia. A regular Japanese magazine, *Fleur*, issue 29 of 1997 was dedicated to proteas with high-quality pictures of plants in nature, in bouquets and arrangements all in Japanese, with not a single word of English.

Industry developments 2000 - 2005
South Africa

It was a good time to be Chairman of SAPPEX. In 2000, I received an invitation to talk at the Florissimo international colloquial in Dijon, France, all expenses paid. The discussion theme was '*Seeking to pool a common economic interest between Producers, Buyers and Distributors of flowers, foliage and plants*'. I centred this talk around the South African perspective on 'Who owns Nature, Certification and Traceability'. I was impressed by the presentation from Chile, and quite jealous about the level of Government support their floriculture gets. If you do some sums, I was quite well paid for my fifteen-minute talk. My fully paid expenses included my airfare to France, the high-speed Thalys train to Dijon, and a three-night stay at a hotel near the venue where a simultaneous flower show took place. The Embassy in Paris was very helpful and I was ferried everywhere by Gordon Gleimius who even showed me some important sights of Paris on the way to the station. To my delight he even came down to Dijon to listen to my talk. In support of the event, Bergflora sent thirty cartons of flowers for a local group to make up a very nice stand in support of my participation. They won an award for the best florist display of a stand.

Proteas had been grown in the North of South Africa, particularly in the Transvaal (Rustenburg) and also in KwaZulu-Natal where there were a number of endemic proteaceae species. SAPPEX and ARC had arranged information days up North, and these areas were now also given attention by publications like *Farmers Weekly*. The members needed to feel part of SAPPEX and we had to do something to be able to attend to their problems and research requirements. SAPPEX therefore started thinking along the lines of regional representation. This clearly brought with it further expense, as the members up there felt that as long as they were paying a levy, they should not have to finance their representative to the Executive Committee of SAPPEX. Questions like 'what is SAPPEX doing for us?' were raised. I get so mad when people ask this question. Do such people ever pause to think what they want SAPPEX to do for them and tell their association about it?

At the AGM that year, Harold Wenk, who with his son had so effectively decorated the SAPPEX stand at IPM, came and did a demonstration of how they arrange proteaceae with traditional garden flowers. The resultant creations were eagerly carried off by members after the event. He brought with him a framed certificate as a token of thanks to SAPPEX for providing such wonderful material for the Silver Rose florist competition in Germany. This was part of our sponsored consignments for floral design schools and floral regional and national competitions.

Oh grief, and then the South African Flower Export Council! When I asked the council if a survey could be done on the floriculture sector to see what we were actually involved in and wanted to take responsibility for, it resulted in a Floriculture Cluster Study — a very expensive, but quite useless report that did not help us in the slightest, but it created quite a stir because of statements like: *being able to increase exports from $30 million in 1998 to $250 million by 2008, with a concurrent jump in job creation*. What I had wanted to know was how many growers and what crops, how many nurseries, how many exporters, the extent of the local market, input costs per floriculture sector, etc. I wanted a 'Malan'-type survey like the one that was done on the Fynbos Industry, for the whole of the country over all floriculture sectors. But the council thought it would be nice to have this study done, and what an expensive disappointment it turned out to be. The people that benefited most from the study were the consultants! Granted there were a few recommendations in terms of government support that would have helped us on the way to their projected goal of floriculture in South Africa, but there were also a number of statements that were way off the mark. As a result of the projections of the cluster study, there was a frenzied interest in Fynbos. Unfortunately, while the government was impressed with the report, they omitted to read the part that said that government support would be required to make these projections a reality. In fact, funding for research and development was under threat, resulting in researchers having to write long project proposals for overseas funding to keep the Fynbos program going.

I had to warn the general public not to climb on the bandwagon in a great hurry, but to take cognisance of the problems still facing the industry, like airfreight space, increasing costs, uncertainly of research funding, and the resignation of researchers due to a shift in focus by the ARC. I was not very popular in some quarters. Publicity about the Floriculture Cluster Study kept cropping up all over the place including Parks & Grounds, the Cape Times, and of course, the *Farmers Weekly* and its Afrikaans equivalent, **Landbouweekblad**.

The Flower Council decided to start a publication for the floriculture sector, called Talking Flowers. As if I didn't have enough to do to publish four journals for SAPPEX and two journals for the IPA, I now had to send them copy for publication. Fortunately, they were happy to receive and publish second-hand news (that I had written up for the SAPPEX journal).

There were various reasons why SAPPEX's relationship with the Flower Council was not a happy one. After a period of time we resigned from the council, and later were asked to rejoin and this was again followed by our resignation. I still think that the Flower Council was not correctly structured/managed but in 2010 SAPPEX and the Flower Council entered into a relationship again. Hopefully, there have been positive steps in the right direction in the intervening years.

Due to lack of funding the Genebank was facing closure. Appeals to the Department of Agriculture seemed to fall on deaf ears. Negotiations with Hawaii to share the costs and benefits of the Genebank did not sit well with a number of the members who did not fully understand that it would ultimately be in their interest. The Floriculture Cluster Study, which had received wide press coverage, had one unexpected positive spin-off — the Fynbos Industry was finally being noticed by those high up in the government and eventually government funding was provided for the maintenance of the Genebank. Fund raising and financial support within South Africa was a very slow process. I had hoped that all those firms using the name 'protea' would be happy to come on board, but no, not at all, which was very disappointing.

A few of the South African exporters tried to follow recommendations of so-called 'fundis' and tried to expand their American market. Small consignments had gone to the U.S.A., entering New York (via Europe) and Miami. I will never forget the phone call I got from our embassy in Washington around then, with the question: 'How is it that Holland exports proteas to New York and South Africa does not?' I told him that our goods were just about blacklisted because of insect and disease problems, and where did he think Holland got proteas from? Although random checks are done on all floriculture imports into the U.S.A., certain problem countries of origin are subject to far more stringent inspection than others.

Anyhow, the firms had gone to considerable expense and trouble to open the American market for South African proteas, with growing success. They were extremely careful to send only 'clean' flowers, meaning varieties that were low risk in respect of harbouring insects. Slowly but surely they were gaining ground, and then South African Airways (SAA) pulled the rug from underneath them and decided to stop the direct flights to Miami in favour of Ft. Lauderdale. SAPPEX of course objected to this, but nobody cared, until I wrote a rather nasty letter to the press, which was published in *Business World*. Suddenly SAA jumped and offered to take me to Ft. Lauderdale to show me first hand how they would look after our interests. I said, 'Fine, but not without Mr. Taljaard, one of the largest exporters to Miami, and if you are taking us for a short visit there and back, please make it business class'. Matie Taljaard and I had to act quickly to get our visas, organise one of his importers to pick us up, and ferry us from Miami to Ft. Lauderdale, as this was not part of the deal.

When we got there, we were taken to a dingy office to talk to the U.S.A. inspectors. We saw nothing of the airport itself, and when I asked about the cold-storage facilities, we heard there were none, and on asking about the flight timetable South Africa, I was told it was due in on Fridays. Anyone who has ever been to Florida knows all about the humidity there and can just imagine what would happen to the flowers between Friday's arrival and Monday's inspection clearance! We did not get to talk to an SAA freight representative, but a person in charge of passenger traffic. Was I hopping mad! What a waste of time! And we were embarrassed towards Matie's customer. In any case, whatever made these jokers think that importers, who are all situated near Miami airport where all the South American flowers arrive, would have either the time or inclination to run to Ft. Lauderdale just for South African products? I must admit that I was not very pleasant to the SAA representative.

There was a sequel to this when the PRO of SAA came to the SAPPEX office a few weeks later to try and appease us. He was clearly very uncomfortable and Willem Verhoogt of Bergflora, who had joined us for this meeting, just shook his head and said, 'I knew it would be a waste of time', grabbed his briefcase, and walked out. You can imagine that our confidence in our national carrier was badly dented. The only positive outcome from this whole adventure was that I got to know at least one importer of South African proteas and that we had a fine dinner together!

On a sad note, Boetie Small, one of the founder members of SAPPEX and regular attendee at meetings, also Executive Committee member for a number of years, died in 2000. At the last meeting he attended he handed over a certificate that had hung in his office for many years. This came from the Pennsylvania Horticultural Society as a thank you for flowers that SAWGRA donated for their Spring Show in 1969.

Another highlight was that South Africa had a national stand at Hortifair in Holland in October 2000. A very smart stand, double story, with a business/entertainment area upstairs and storage below for coats, hats, buckets, etc. The 'roof' was supposed to be a *Protea cynaroides*, but to me it looked like American-Indian headpiece of feathers! A number of SAPPEX members had a section on the stand, while the SAPPEX side was manned by Hans Hettasch and me.

South African stand wins the coveted Csezik trophy at Hortifair, Amsterdam, 2000 (FloraCulture International Jan 2001)

To build up the stand was quite a madhouse, but we were ably assisted by Heidi van Rensburg and Renita Swart of Honingklip Dryflowers, who were to meet Robert on a business trip to his clients all over Europe a few days later. The South African stand won the prestigious Csezik Floating Trophy and the SAPPEX section appeared on the front cover of FloraCulture International. This of course raised the profile of SAPPEX and proteas and slowly but surely we started getting more recognition in FloraCulture International. In May 2001, they published an article on our promotion efforts in Germany.

In 2002, a group of forty-two German Florists from Rhine-Westfalia visited the Cape and were shown around the Cape, with the obligatory cable car trip to the top of Table Mountain, followed by a visit to Hanzel and Merlette van Zyl in Stellenbosch. In the late afternoon, they went to the wholesale premises of Pieter de Bruyn and his partner and met up with South African florists and embarked on a flower arranging workshop, with proteas and garden flowers. Mr. Vogelpoth, their Chairman was very keen for Pieter to come to the IPM in Essen, Germany to give a floral demonstration there. Shortly after that proteas were given the front page of Floriculture and were used in photo-shoots and special articles. What a great success that whole campaign was! Tiel gave numerous talks to SAPPEX members on the success of our participation at IPM that he used as a springboard for the promotion programme. I think to this day not many people appreciate how much they owe Tiel Bluhm for this achievement.

In September 2000, proteas again made it to the front page of the *Landbouweekblad*, together with an article on the potential of the Fynbos Industry.

SAPPEX involvement at Fynbos Forum continued. At the 2000 Conference, I gave a short talk on the establishment of Honingklip's Private Nature Reserve, situated as it is in the buffer zone of the Kogelberg Biosphere Reserve. Fynbos supplied us with a living and this was a way to show our appreciation by conserving some of this valuable but threatened ecosystem. Richard Cowling decided it warranted an article in Veld and Flora, where it appeared in the March 2001 issue.

SAPPEX and the researchers held a joint research meeting where Dr. Gerhard Jacobs outlined where money had been spent the last few years,

Tiel Bluhm

and particularly the post-harvest work was showing good results. Eventually, the knowledge gained resulted in certain Fynbos species being shipped in reefer containers (temperature-controlled containers) to Europe, but this had to be carefully orchestrated so that the market is not flooded at any given time. The colour research leaflets continued. By 2001, the industry had invested close to R1 million in research and publications over a period of four years, all made possible by the voluntary levy and hard work.

In 2001, as a result of the large amount of Critical Ecosystem Partnership Funding (via the World Bank) that came in for research and social programs in the Cape Floral Kingdom, an organisation called CAPE (Cape Action for People and the Environment) was born to act as a conduit and selection body. They became an accepted part of Fynbos Forum where they first came to introduce themselves to explain the concept of working through them for funding allocation for deserving projects.

SAPPEX, in partnership with ARC, had by now become a regular feature at the only real commercial flower trade show in South Africa, called Gardenex. In 2001, we set up our own stand — how could the only indigenous South African flower industry not participate? As always we had wonderful collaboration from various members and this time we had the cooperation of the Summer Rainfall Study Group under the leadership of Gordon Bredenkamp of Jupiter nursery. Our stand was professionally set up by Jannie Claassen whose work I had seen on the South African stand at Hortifair. We were very pleased to receive a silver medal.

The protea promotion campaign in Germany was now in its third year and was gaining momentum. It even led to interest from other countries. On the Swedish National Day, we were asked by the Royal Flower Foundation to provide flowers. The TV program on Princess Royal Victoria's charity fund and other activities gave good cover on our flowers.

Meanwhile, SAPPEX identified a number of key issues that needed attention. Cost-saving ways were discussed with regards to distribution of the journal, with thoughts of two journals per annum and monthly email newsletters. Also thought was to be given to a better way to release cuttings to the industry. Producers were surveyed regarding their priorities, but in spite of a lucky draw of a ticket to Hortifair in Holland, very few questionnaires were returned. Then SAPPEX undertook a road show to talk to producers to find out why they were so dissatisfied, with the aim of starting a Producer Committee. Representatives were chosen at regional meetings to serve on the Producer Committee with a delegate to the main committee. This seriously increased expenditure for air tickets and car hire for those who came from Gauteng and KZN, and reimbursement for petrol expenses for those who travelled from the Eastern Cape. The idea was good, but the producers at large were still not satisfied! However, some good came out of the Producers Committee. In the following year, they initiated a growers' survey listing what was planted and where, and this information was correlated by Carlo Pieterse. This was an important tool for producers and exporters.

The Flower Council visited the U.S.A. on a fact-finding mission organised by the Department of Trade and Industry for various commodities. Keith Brooke-Sumner represented SAPPEX and was accompanied by Elaine Reekie, the KZN representative of the Flower Council. It was obvious from his later report that it would require big bucks to market successfully in the U.S.A. where customers require a direct presence, and where it was thought that it would be ideal to establish a permanent South African flower office. Unfortunately, this was impossible as it needed far more money than our small industry, or even the Flower Council could bear, and there was yet a further problem with airfreight, with the flights to Ft. Lauderdale, bad as they were, having stopped in favour of Atlanta.

In 2002, SAPPEX again had issues with SAFIC, the S.A. Flower Industry Council. None of the matters SAPPEX brought to the table were given any attention. For instance, airfreight was a serious issue but did not get any attention in spite of numerous requests. Prompted by our resignation, the National Agricultural Marketing Council, who was represented on SAFIC, ordered a Section 7 investigation into Airfreight problems.

A protea again made it on the front cover of *Landbouweekblad*, 16 August. This publication featured an article on the new policy of the ARC regarding cultivars and cultivar registration after collaboration with the cultivar committee consisting of SAPPEX and ARC representatives. The days of unilateral decisions were over.

The Fynbos Forum kept growing, with over 150 members attending the meeting held in Calitzdorp. For the first time, the committee (on which I also served) had to run parallel sessions. The hot baths were quite an attraction and let it be known that on the second morning of the conference there were a few very tired people, and that Calitzdorp port and a hot pool in the moonlight had something to do with it. A field trip to the Gamkaberg research under the leadership of Jan Vlok was highly interesting. It was quite unexpected to find so much Fynbos on top of some of the Karoo koppies.

At the end of 2001, our first email Newsletter was sent out to members. In this newsletter, we announced the passing of Joan Pare on 18 October. Joan was a founder member of SAPPEX and our ambassador extra-ordinaire having promoted proteas locally and around the globe throughout her adult life. Her dedication had been recognised when SAPPEX awarded her Honorary Membership. Australians will remember her as the person who decorated each and every vase, nook, and cranny at the IPA Conference hotel in Perth, where even the porcelain pig between floors sported a garland of proteas!

We also remembered the horror of 11 September, which had such an impact on everyone's lives.

A new generation consisting mainly of young and professional growers started to take centre stage within the Executive committee. Zac Isaac of the Executive Committee had been delegated to go to Hortifair on SAPPEX's behalf and in his report said that he learnt a lot about the floriculture industry. He felt that we should continue to participate if we could. SAPPEX published a well thought out document to present to government and opinion makers, entitled 'A Case for Growth' in order to raise support for our industry. First details about EuropGap were published to familiarise members with the necessity of getting used to social, environmental, and product-standard requirements.

Our promotion campaign was receiving such good press that a Dutch film group came to make a documentary on South African flowers, which would form one of an eight part series. Bridget Baker, my brand-new assistant to help me with the volume of work, took them around to see the Gibsons, the Middelmanns, Morgenster, and then to Cino Gironi and the Dutch Flower Group. Walking them through the veld, one asked, 'Did you plant all this?' and jaws dropped when I said, 'No, this was planted by Mother Nature'.

A new development in the industry was the formation of an independent company formed to do their own breeding work. Three South African growers, in different areas, plus an exporter and an importer decided to collaborate and fund this company. This was a sure sign that the industry was now coming of age. It is only through such collaboration and progressive action that the flower industry can go forward while ensuring that cultivar material stays in the hands of the individual growers and their particular marketing chain.

To the best of my knowledge, SAPPEX had to take legal advice only once. One of our members, who will remain nameless, was adamant that an additional charge be added to an 'interim' membership increase, was unconstitutional, and threatened legal proceedings. The upshot was a letter to the member concerned, pointing out that he was in error, after which we heard nothing more. By the time I left SAPPEX, he still had not taken up membership again. I saw him subsequently at an IPA Conference and our meeting was quite cordial. I am just wondering whether I have missed my vocation, because the lawyer I consulted first agreed with the member, until I pointed out the error of his ways! It was quite a challenge, but it turned out that it was just a little practise exercise for worse to come.

In September 2003, I hit the most challenging month in my life when an article appeared in 'Africa Birds and Birding' that protea farmers were responsible for poisoning sugarbirds. This article appeared while I was at a Fynbos Forum meeting in Hartenbosch. Fortunately, I had lots of friends there who advised me on what course of action to take on this matter. To see my small information stand adorned by a small wooden cross with a picture of a sugarbird and the letters RIP upset me terribly. That the perpetrator of this churlish action did not have the courage to come up to me directly is still something that burns me up when I think of it. Immediately on coming home, I set the wheels in motion to personally interview each and every member of SAPPEX, a duty shared between me and the Secretary, Margaret Rabie. A furore had started in the press and SAPPEX even received threats of a boycott via an international birding web site. I had promised a thorough investigation, which took a month of concerted effort; everything else was put on the back burner. Results of the investigation were released to the South African Birding Association, the press, and in SAPPEX News. So were any members guilty? There was one member who, when he had purchased the farm a number of years ago, was made aware that the staff were killing sugarbirds on instruction of the previous owner, but he had put an immediate stop to this. There was a suspicion of a grower in the Langkloof, but all investigations lead to a dead end or nobody was talking and nothing could be proved. Fortuitously the alleged Langkloof farmer was not a member of SAPPEX. If there was guilt, I think the person got a bad (good!) fright and would not ever do this again! Fortunately, SAPPEX came out of this with its good reputation intact (SAPPEX News 117, Oct 2003). At a subsequent IPA Conference, I gave a short talk on just how careful one has to be when talking to a journalist!

We decided it was time to meet up with some members in the Southern Cape and for once to travel to Zac instead of him having to drive all the way to Cape Town. We therefore arranged for an Executive Committee meeting to be held at Frans and Meg Gerber of Forest Ferns near Storms River. A number of members joined us socially after the meeting. It really was good to see the up-country members again and Zac did not have the arduous drive all the way from Plettenberg Bay. Frans had become acquainted with the protea industry when he was Operations Manager at Perishable Cargo Agents in Port Elizabeth. He resigned in 1985 after having acquired a lovely piece of ground in the Tzitzikama on the Garden Route, which he in time developed into a major Ferns and Greens operation. The Gerbers have really carved out a superb spot and their accommodation is top class, overlooking a waterfall, while the Fernery is worth a visit. Their farm forms part of the three-day Dolphin Trail, which starts at the Storms River (in the opposite direction of the Otter Trail). A wonderful trail where someone else carries your personal goods and all the food is brought to you along the trail. All you have to do is to concentrate on the scenic route that takes you from the ocean, through the Fynbos into the indigenous forest and back again a number of times. The three nights along the way were spent in absolute luxury. I had first-hand experience as a sixtieth-birthday treat!

In South Africa, all kinds of laws were being upgraded and various levies were instituted to improve the skills of uneducated people. Black Economic Empowerment (BEE) policies, encouraging businesses to promote small business initiatives from previously disadvantaged communities were being written into government procurement rules and if you wanted funding from the government, you had better be sure that you had a BEE component. Labour laws, security of tenure on farms and all kinds of things were changing that needed a new way of thinking on how we do business.

Suddenly environmental concerns were getting priority attention and we learnt more about biological control and more nature friendly ways of farming. Certain 'alien' products that had been utilised by the dried flower exporters could no longer be traded. SAPPEX stepped in and called a meeting with Nature Conservation officials to discuss the pros and cons. A new set of rules were drawn up for implementation by the dried flower fraternity. Certification programs also needed far more attention to ensure that we could continue to export to the European market where flower labels and traceability were becoming more important for trade. Certain pesticides were to be phased out by Europe and all these things meant that farmers had to keep up to date with latest developments. SAPPEX did its best to keep its members informed.

During 2003, a Chinese delegation from the Department of Agriculture visited Honingklip Farm on IPA business. They had already seen the Australian protea growers and now they wanted to find out more at the source of our genera. They were very complimentary about what they had learnt and seen at the farm. It would be very interesting to see how their protea plantations are coming along. A visit to China as a tourist is definitely still on my bucket list.

At the end of 2003, there were rumblings of great dissatisfaction and underlying problems in the industry. It seemed to me the more we did for the members, the more dissatisfied they became. By then, in my opinion we had reached a level of professionalism that provided real value to the producer. The problem of course was that it all cost money and a group of members did not think that the voluntary levy to which they contributed, gave them, individually, value for money. Malcolm Wallace of the Executive Committee was tasked to facilitate meetings with members to determine the wishes of the members as to the role and functions of SAPPEX. At a subsequent think tank, everybody agreed that the Protea Industry needed a unique organisation and that it needed promotion and research. The objections therefore came from people who were not willing to attend meetings, and who probably did not really read the journal or newsletters. Some thought that SAPPEX should facilitate marketing, which was clearly outside its aims and objectives. So, the outcome of the think tank was that the role of SAPPEX should be more clearly defined.

There is no doubt that the SAPPEX Executive did a wonderful job in maintaining research capacity and feedback of results, publication of research leaflets, promotion and information sharing. And so we continued!

In October 2003, we had to say farewell to another well-known personality in the Fynbos Industry, Jimmy Yeats, who with his wife Carol had first run a protea farm, and later expanded into an export business from their base in Hermanus. After Jimmy and Carol retired, their daughter Jamie rekindled her interest in proteas and participated in a good number of SAPPEX events.

2004 arrived and with it the news that some exporters were no longer willing to pay the levy to support research and promotions. We were fully aware that this would mean that more people would follow their example. This was really bad news! In order to start cutting costs, we decided to start charging for additional services. Therefore research leaflets, back issues, and research papers would have a small price tag attached to them to offset the costs. The producers committee was still trying to ascertain what the gripes of the growers were all about.

The expected crash in the levy system resulted in SAPPEX investigating Statutory Levies. In this I approached Godfrey Rathogwa of the National Agricultural Marketing Council and other leaders of agricultural sectors in the Western Cape. The committee came to the conclusion that, considering that even a statutory levy (which needs approval from the majority of members) requires policing by the industry for whom it is instituted and that the extensive reporting required would create another administrative burden, that we would leave matters as they were.

SAPPEX applied for, and received funding from the Western Cape Department of Economics to enable us to continue our promotions in Germany. We also received funding for appointing a consultant to assist to do an investigation into a Fynbos-specific basic certification program. After months of effort, we took the document around the country for discussion and input by the members. Then it was launched at Gardenex and presented to the Flower Council. They were told that this could be adapted to any other floriculture sector, as long as they acknowledged the source. We had even designed a special logo for the certification program, which contained the slogan: *'concerned for people and the environment'*. To the best of my knowledge Barrie Gibson was the first protea grower to be EurepGap certified. So, here was an opportunity for more SAPPEX members to start a basic certification program.

The year 2004 delivered a major highlight when our South African florists Andre Louw and Pieter de Bruyn did a stage presentation at IPM Essen with the internationally renowned floral designer, Florian Seyd as moderator. They made spectacular arrangements with proteas, and used typical South African decoration products like ostrich eggs, succulents, dried cones, porcupine quills, and much more. They chose South African penny whistles and the Soweto String Quartet as background music. The stage presentation is the undisputed centre of all florist activities at IPM and is considered by many as the centre of 'world floristry' as far as trends, skills standards, and presentation are concerned. All this came as a result of Tiel Bluhm's hard work and effort over the last few years. It needs to be told that the airline had delayed Andre Louw's luggage and that he arrive in Europe in the middle of winter, with only a light shirt and pair of pants. An urgent shopping trip was needed to get him kitted out for the extreme cold. Not a very nice welcome for a first-time visitor!

IPM Germany, demonstration. L to R: Tiel Bluhm, Pieter de Bruin, Florian Seyd, Andre Louw with IPM organisers.

The November edition of *florist* carried a four-page feature on proteas written by Andreas von der Beeck of Muenster. The accompanying pictures showed arrangements of proteas mixed with other flowers for year-round-use, just what we were trying to achieve.

In 2004, Jeudi Petersen joined as PA to the Chairman, replacing Bridget who had left for the UK. A long-awaited book on diseases written by leading pathologists was launched. This mammoth job was done by South African's Pedro Crous, Sandra Denman, Joanne Taylor, and Lizeth Swart in collaboration with Mary Palm of the U.S.A. It will remain a valuable textbook and reference work for years to come.

SAPPEX submitted a project on Sustainable Harvesting to the Critical Ecosystem Partnership Fund. This was aimed to bring the conservation message to everyone who utilises the floral resources of the Cape Floral Kingdom and to promote involvement in conservation. I had collaboration with a number of high profile people and the result was a useful little booklet of which I was really proud. Ann and Mike Scott took the booklet and certificates of attendance on a road show throughout the Fynbos regions in 2005 and numerous CDs were also produced and distributed to Nature Conservation offices as study and reference material. Again it raised SAPPEX's profile as a responsible organisation. The African Grant Director, Nina Marshall, wrote in her letter of acceptance of the project results: *'All programmatic reports have been received and approved. I would like to congratulate you on your fantastic publication. 'An Introduction to Sustainable Harvesting of some Commercially Utilized Indigenous Plant Species in the Cape Floristic Region'. This booklet is an excellent resource — packed with useful information and instruction, illustrative photos, and in two languages no less. You really have outdone yourselves during this project!'*

CAPE certificate to SAPPEX for Developing guidelines for sustainable utilization of Fynbos, 7 June 2005

Then I was invited to a meeting of some producers, who advocated the formation of a Section 21 Company to act on behalf of all protea growers. I tried my best to make them understand that they could just as well do this as a subcommittee of SAPPEX. They could be authorised to manage their own affairs, under the SAPPEX structure and even take the responsibility of managing the research fund. Unfortunately, they were determined to follow their chosen path. SAPPEX would have to make some drastic adjustments.

Of course life carried on and we had a nice little stand at the Cape Town flower show, decorated with help from Pieter de Bruyn. It drew quite a bit of interest and it was good having some members there with their own stand. Hanekom's Caledon Fynbos Nursery and Flower Valley were in close proximity to the SAPPEX stand.

In 2005, we published our accreditation logo and we celebrated the fact that Birdlife South Africa was entirely satisfied with SAPPEX's report on the Cape sugarbird poisoning allegations. The Section 7 investigation into the South African cut flower industry was received and while it was not Fynbos specific, it had a number of valid recommendations with action to be taken by the industry with support from various government departments.

For SAPPEX's fortieth birthday, which was held at Devon Valley Hotel in Stellenbosch, I had prepared an overview of forty years of SAPPEX. It was good fun to dig back into old newsletters and journals and pick out the highlights. We organised a special dinner and succeeded in keeping costs down through sponsorships. During the evening we announced the first persons who had received the Fynbos Certification Certificate. SAPPEX also published a commemorative issue of its journal covering forty years. During the afternoon's AGM Peter Dorrington advised members of the newly established Section 21 Company, Protea Producers of South Africa (PPSA) who would be handling all producer-related affairs in the future, while SAPPEX would remain as a body that deals with legislation and that it would restructure in such a way that it would satisfy the partnership between industry sectors. The secretariat would be given an opportunity to re-focus and change the way in which SAPPEX would function. The producer committee would therefore no longer fall under SAPPEX, but a producer would be delegated to the Executive Committee.

At the Fynbos Forum in Mykonos, Langebaan, four people were honoured for their exceptional contribution towards recognition of the Cape Floral Kingdom as a World Heritage Site, namely James Jackelman, Lee Jones, Paul Britton and Guy Palmer. Each one was a deserving recipient. Having reached such an important status of course increases the onus of Fynbos landowners to really participate in conservation.

National Botanic Institute, Kirstenbosch, announced that it would no longer export seeds or distribute same to overseas members, due to the high risk of causing invasive problems for importing countries. At the research centre at Kirstenbosch a garden featuring 'Invasive plants we have given the world' was established. Lovely plants from our point of view, but not always so nice elsewhere!

International Developments

International Protea Association

The tenth International Protea Association Conference and fifth International Protea Working Group Symposium took place in Puerto de la Cruz in Tenerife in 2000. In spite of the distant destination, protea growers and researchers from all over the world came to be updated on the latest developments and to see yet another protea-growing region. A few of the delegates got left behind in Cape Town due to a mess up by SAA, and arrived late. Instead of going via Madrid, they had to travel in different directions to eventually reach Tenerife! (Shades of the Amazing Race!) Poor Conrad Archer from Zimbabwe arrived, sans luggage and had to borrow a pair of shorts . . . !

Although protea growing is difficult in Tenerife and Madeira, and the farms are relatively small, people are hardworking and keen to make a success of it. The conference itself was stimulating and the cultural program combined with the scenery and natural environment was really something special for the visitors. A good time was had by all and that includes Maryke riding a pink elephant! (If you weren't there, don't ask!) Juan Rodriques Perez, President of the meeting, was a wonderful host and he and his team could be justifiably proud of their efforts. I am still in awe when I think back to the person who did the simultaneous translation into Spanish for the local participants. That must have taken quite some doing.

At the conference there was talk of international collaboration for a donation of flowers for the AIFD National Symposium in the U.S.A., which eventually resulted in flowers being donated from all around the globe, supported by a small amount of funding from the IPA. Apparently the event was very successful and the proteas got good international exposure.

On the international front, we were delighted with the lovely Australian indigenous bouquets presented at the Sydney Olympic Games.

Phil Parvin, that irrepressible godfather of the Protea Industry who had retired back to Florida from whence he had come to take up his post in Hawaii, managed to involve himself in a new 'old' interest. It might surprise some of you that he joined the Charlotte County Cultural Centre Community Choir. Not surprising is that he took over the job of Secretary for the Rotary Club of Port Charlotte. Partner Ron keeps himself out of mischief with a voluntary job to manage a food pantry for HIV/AIDS patients. During a visit to these good folk we were recipients of Ron's great cooking skills. No wonder Phil put on a few odd kilos. Fortunately, email keeps Phil in touch with his protea friends in Hawaii and around the world. We managed to entice him to join us for the IPA conference in Hawaii on 2002, where he shared his extensive experiences of protea cultivation around the world, with an overview of the past and present, with challenges for the future.

What a treat to be in Hawaii again for the 2002 IPA Conference on Maui, and to catch up with friends made there over a period of time. In terms of sheer luxury, nothing could beat our conference hotel, the Renaissance Wailea, with its tropical garden going right down to the beach. And what a beach, what a sea: white sand, gentle waves, and the most incredible snorkelling within a stone's throw from land. Early morning I had to have my swim and it was wonderful to be under water and hear the whales miles offshore. I experienced a real first for me, and that was to swim with large turtles. The social side was great, the food was out of this world and of course the conference itself was top-notch as always. The Luau feast was something unique, but even better was young Mr. Lawrence Kellar, ex South African, on stage with the girls, being taught how to do the hula. Imagine

Conference logo

this bulky guy, more of a rugby player type, trying to get that sexy hip-sway right. We laughed till our belly's ached. On the other hand, Katy Percival looked like she had done it since birth! Another outstanding feature of the conference was the amazing conference design, an aboriginal artwork incorporating pincushions (Leucospermum) on the four corners. These were put on bags, stationery, t-shirts, and cards. We enticed Phil Parvin to come to Hawaii where he was honoured by having a deep red *Leucospermum* called after him, Ls. *'Phil Parvin'* during the conference.

Happy participants at the opening function — fltr : Conrad Archer (Zimbabwe) Dr. Rodriques Peréz (Tenerife) and Geoff Jewel (NZ)

At the IPA Hawaii closing dinner pit-BBQ, fltr: Robert Middelmann, Margaret Rabie (Secretary/Treasurer), Lawrence Kellar (California) Maryke Middelmann (Chairman) and Cecily Saywood (Zimbabwe)

I forget the reason, but South Africa's Gail Littlejohn, highly respected by scientists and growers around the world, was prohibited by the South African Agricultural Research Council from attending the conference. No matter how much we objected, they would not see reason, and Gail's four-year EU-funded research had to be presented by others. She of course could not discharge her duties as Chairman of the Research Committee. We really missed her presence.

Australia would be the host for 2004, with Alison George as President. The conference would run concurrently with the Melbourne International Flower and Garden show in the historical Melbourne Exhibition Building. We kicked off with a wonderful pre-conference tour from Sydney to Melbourne. There was so much to see, and so much fun was had by all that it really was a great working holiday. Apart from the protea events, highlights for me were the sighting of Koalas and the miniature penguins on parade near Philips Island. A post-conference tour taking in a number of protea farms was equally successful. Walter Middelmann wrote a special note to the IPA for the conference:

> *Having been present among the delegates attending the first Founding Meeting of the IPA, also at Melbourne, my thoughts are with you but, at now age ninety-four years, I do not feel up to undertaking the journey nor do justice to its program and purpose.*

An interesting feature was that we held the General Meeting at a dinner venue at Clover Cottage in Blerick. This was a charming venue and I had imagined that it would be very difficult in such a setting, but on the contrary, it went very well. Rita Mathews honoured us with her presence there. It was also at this event that David Mathews announced that Proteaflora would make a sizeable grant for research in memory of Peter Mathews. This fund is now called the Peter Mathews Research Fund and it was decided to only use the interest and the IPA would try to grow the capital sum.

The next conference would be held in San Diego in March/April 2006.

International collaboration continued with the exchange of newsletters and journals to the IPA. Over time the content and layout of all these newsletters and journals improved; without a doubt the new computer programs made it a lot easier for everyone involved to produce better looking and easier to read text. *Protea Press*, the quarterly magazine of the New Zealand Protea and Foliage Growers Association is a prime example of a well-presented journal.

My predictions to the IPA in Melbourne that the South African industry would undergo some major changes came true sooner than I thought!

California

Howard Asper is recognised as the person who introduced the South African protea to America. He brought out protea seed from South Africa in 1961 and propagated it successfully in 1965. He died in 1993, but posthumously was inducted into the US Hall of Fame, for the sixty-six-year career during which he made numerous valuable contributions to the Californian floral industry.

Following on the Hawaii IPA Conference, there was international collaboration for an American Institute of Floral Designers (AIFD) National Symposium under the theme 'Ancient Echoes — a Future Wave' where with dramatic sounds of thunder and lightning on a dark stage, fifteen superb protea arrangements were introduced. Dean Yale of the AIFD received a standing ovation for this presentation and the press made a mad rush to take photos after the arrangements were moved to the hallway after the presentation. How different his report was to that of Pieter de Bruyn who was also in the audience. Pieter attended because he was inducted to the American Institute of Floral Designers at the symposium — only the second South African to have achieved this distinction. Pieter was very critical of the program and the fact that none of the new cultivars were mentioned. Dean Yale referred to Brunia as 'these silver baubles' and said to Pieter that he wasn't at all worried about the names of the proteas. Pieter felt that the protea products sent to San Diego were not well promoted and advertised. This was contrary to Pieter's glowing report of collaboration for the World Flower Council Summit in Amsterdam, where OZ Flora arranged for delivery of Proteas that were used at the Summit and at the Floriade. Pieter assisted Lea Liebenberg AIFD, and Jeanette Pugh, designer. A number of delegates from other countries also used proteas at this event. His closing words of the report were 'together we learn from each other, we share, we laugh . . . what a wonderful world we live in'.

Australia

'Sugar & Spice', a *Serruria florida x barbigera* was launched by Proteaflora. Also in Australia a guide to the business of joining the Australian Protea Growers Association was published under the name 'The growing business'. Australian Flower Exports and Renewable Resources, both situated in Western Australia published some very nice pamphlets for their businesses.

Netherland

It was with great sadness that we heard that Klaas van Zijverden, one of the pioneers in protea imports into the Netherlands, had passed away. He was not only a well-liked and internationally respected figure in the floriculture business, but he will be remembered for his leadership and commitment to his community, for which he received a Knighthood. He was a very special man, who will be remembered with respect by the business community and affection by those who knew him.

2005 Onwards
The Break-up of SAPPEX

The formation of the Protea Producers of South Africa (PPSA) and their unwillingness to be normal members of SAPPEX was the reason for putting together a task team to look into the way SAPPEX should function. Neither the dried flower exporters nor the fresh flower exporters were keen to start up their own Section 21 companies to resort under the SAPPEX umbrella. PPSA had made it quite clear that they needed a SAPPEX as umbrella body, and that each sector should delegate a member to the Executive Committee of SAPPEX. In order to find the best outcome, Maryke, Margaret, and Jeudi were required to do a job-and-responsibility analysis to see how costs could be trimmed now that PPSA had taken over the producer part of the levy and now that the exporter's levy was under threat due to the withdrawal from this levy by one or two exporters. All this happened at a time when SAPPEX really had its hands full to not just serve its members, but also to maintain good relationships with many other organisations and government bodies. It was my firm belief that SAPPEX could not operate properly in a vacuum and that this interaction with other organisations could only benefit SAPPEX and all of its members.

Certainly the future looked quite bleak and I for one was working often till far beyond normal office hours to cope with the volumes of work. I had already informed the Executive Committee that I had appointed Jeudi to take over the everyday running, while I would be quite prepared to continue as Chairman. Costs would not increase as I would no longer feel entitled to be paid for my secretarial services that I had provided for so many years in addition to the duties as Chairman.

The PPSA required a cost estimate for SAPPEX to handle their secretarial work. This and other in-depth studies followed months and months of negotiations and discussions. As time progressed the situation became more and more precarious and it was all quite difficult as alternative ideas about the future of SAPPEX came thick and fast. It became increasingly difficult to do the job and not to get despondent about the direction things were going. With the membership having shrunk to about half, the bookkeeping and membership functions would no longer need a full-time person and all this had to be thoroughly thought through. We were even subjected to a workflow analysis consultant to see where we could streamline further. The only thing she could come up with was that we should start to charge more for information generated for members. So I guess we had not been doing too badly on our own.

Added to this was the uncertainty of the Fynbos Genebank and the Genebank Trust, which, as it turned out, had never been registered by the previous ARC incumbent. That meant that there were no real 'Trustees' and with the resignation of Fienie Niederwieser and Emmy Reinten, nobody was left at ARC with any knowledge of the 'Trust'. The 'Trustees' therefore were only an interest group and our only external 'Trustee' also wished to resign due to this precarious situation. We had to think of disbanding and distributing the funds.

Peter Dorrington was the main player of the newly formed PPSA and there were quite a number of discussions between us. He was adamant that their aim was to build a strong producer body, but that SAPPEX must remain as the bridging association. It was hard to accept as we could see a mile off that fragmentation would eventually lead to the demise of SAPPEX's strength. (I recently came across a memo from HLD Wood, written in 1980 addressed to the Executive Committee of the time. His statement is still equally valid today. He wrote: *'I must stress that an Association of this nature, if it is to work at all for the good of the Industry, must have the fullest support and cooperation of all aspects of the industry and its members. Without this it will never work and it would be far better to dissolve it altogether, as soon as possible'*.)

PPSA agreed to delegate a member to the Executive Committee. The current Producer Committee, which was a body constituted by SAPPEX, was supposed to meet before the next AGM to take a decision on the way forward, but it was clear that SAPPEX could no longer finance their meetings. I was getting increasingly worried about Margaret's position and I was equally upset about the uncertainty of Jeudi's future in SAPPEX. She had fast become a real asset, in spite of having to cope with a private tragedy at the time.

Meanwhile, I continued my normal activities on behalf of SAPPEX. Once again I had a motion to table at the Agri-Western Cape Conference. This time I felt it necessary to point out that the lack of management at the Agricultural Research Council was having a very serious negative impact on the Protea Industry, and called on key role players to lobby for change in the way the ARC was run. Loss of research personnel was a huge cause of concern and the industry had no alternative but to increasingly swing its research support to universities who could deliver results in a more cost-efficient manner.

All the while in the background and after two years of uncertainty and discussion, it was clear that a lack of financial support for SAPPEX, resulting from the formation of the PPSA, and withdrawal of the levies from the fresh flower exporters, would reduce the effectiveness of the association to such an extent that it would be hardly worthwhile to continue. Under the circumstances, the Executive Committee had little option but to consider dissolving the association. The PPSA were of the opinion that they could do better. On 9 February 2006, a Special General Meeting of SAPPEX was called where the existing executive co-opted the PPSA Board plus Willem Verhoogt, after which the present incumbents tendered their resignation/retirement. For me it ended twenty-five years of being involved in SAPPEX; first as Secretary, then four years later as Committee member and for the last seventeen years as Chairman.

At the end of February I received an email from Tiel Bluhm that says it all:

I am really not up to date yet with the latest developments at and its exact consequences. I am sorry that the whole thing seems to have broken apart this way. Thanks for all your efforts over the years. I am sure it has been appreciated by many members. It's just sad that it all ended in such an unappreciative, inconsidered (sic) and ruthless way.

The new committee's first action was to have a facilitated strategy planning session to establish what is needed from the industry representative body. And then an action plan would be put in place to restructure the industry body to function as a mouthpiece of the entire industry and be fully representative. Nothing very new came out of that meeting.

As a last service to the industry, I took action after a devastating fire destroyed approximately 46,000 ha of Fynbos in the Overberg in February, threatening 4,000 livelihoods. I approached the then MEC for Agriculture, Mr. Cobus Dowry and explained to him that Fynbos farmers could not ensure their crops, in spite of SAPPEX's years of discussion with various insurance companies. Therefore, these farmers would need urgent assistance. He responded by calling a meeting with banking institutions who agreed to reschedule farmer loans. He furthermore applied for treasury funds for poverty alleviation of workers who were now jobless. This led to an extensive 'working for water' programme being initiated to destroy regenerating alien plants that were such a huge threat in the area.

At the end of 2006, the German trade magazine *Florist* came out with a wonderful feature on the Cape Flora, filling pages with 'exotic' proteas together with Poinsettias (or Christmas Stars as they are known there) that are so well established for the Christmas trade. It was the culmination of many years of collaboration and hard work, particularly by an enthusiastic and knowledgeable Tiel Bluhm, who had also had the rug pulled from under his feet by the changes in SAPPEX. If ever there was a success story, this was it, and I wonder if the industry will ever have such an opportunity again. But, perhaps Germany has had enough attention, and perhaps in collaboration with the main importers the industry should look further afield for a new promotion effort.

In December 2006, Walter Middelmann, founder Chairman of SAPPEX who together with his wife, Ruth, initiated of a number of Fynbos commercial activities, died at the age of 96. Brian Huntley was kind enough to offer us the use of facilities at Kirstenbosch for the celebration of his long, varied, and inspiring life. This event was well attended by friends from all walks of life, the Fynbos Industry, and representatives from many of the organisations he had belonged to. I think he would have approved of this celebration.

At the next SAPPEX Annual General Meeting, the following July, Peter Dorrington made a point of stating how much money would be saved by closing 'three offices' and outsourcing to an office in Paarl and by stopping all non-core activities. Such gross distortion of the facts was highly unfair. This is however, not the right place to set the record straight. But I feel that I must state that it was a sad note on which to end to my involvement. A little ray of light came from an unexpected source. Rudi Visser walked into my office one day with a lovely colour drawing of proteas, as a token of thanks from Floraland for everything I had done for the industry. It hangs here in my home office and I enjoy seeing it.

Over the next few years, the PPSA struggled to find answers on how to keep SAPPEX functioning as a representative body, and it seemed that SAPPEX was dying a slow death. Some sterling work was done by some of the individuals in charge. Gerrit Nieuwoudt, just recently having returned to the protea industry, was one of them. In spite of their hopes to encourage and succeed in forming different Section 21 Companies for the different sectors, there was not much cooperation from the members. Export statistics were no longer gathered. It was particularly sad to see that the total kilogram exported per annum, which had been meticulously kept from 1976 (1,024,986 kg) to 2005 (4,572,068 kg) were no longer recorded. The Protea Industry turnover (dried and fresh) was at that stage valued at R212 million. The journal was replaced by sporadic 1 pagers and the PPSA itself was not really growing. A lot of work was done behind the scenes, but the general membership was not really informed of what was happening and therefore interest in the affairs of SAPPEX was rather low.

At each subsequent AGM, promises of progress were made, but apart from some whitewash, nothing really resulted in the ensuing years. Granted, the PPSA committee had committed themselves to running the next IPA Conference (2008), which kept them very busy. Unfortunately, and in spite of best efforts by a number of people, by 2009 the industry was more fragmented and less representative than ever. The fact that there were only sporadic newsletters did not encourage a strong and vibrant membership. Good news was that PPSA had negotiated via a Dutch importer to access research funding from the Netherlands via the Productschap Tuinbouw. SAPPEX has again joined SAFEC, the Flower Council, and I can only hope that the latter organisation has improved over the years and that it will be a more fruitful relationship.

According to a report in the Afrikaans agricultural magazine, **Landbouweekblad**, 90 per cent of production of Fynbos is still occurring in the Western Cape, and of the estimated value of the Industry, pegged at R300 million, 90 per cent is from plantings with only 10 per cent from veld harvested products. Producers have decreased to 150, but their individual acreage has increased to an average of thirty hectares. The ARC Fynbos research unit, after a long period of uncertainty, moved from Elsenburg and is in the process of upgrading their infrastructure so that they can work on Fynbos again. The Genebank 'Trust' was finally wound up and the monies distributed, mainly to the Agricultural Research Council for re-establishing the Genebank at Nietvoorbij, their new premises. Awards from the 'trust' were also made to the IPA for the Peter Matthews Research Fund and to SAPPEX.

In December 2008, Peter Dorrington retired from office. Denis Shaw, a long-standing SAPPEX member who at various times has served on the Executive Committee also as Vice-Chairman, was asked to assist. Moves are now afoot to re-establish SAPPEX as a strong body to represent the Fynbos Industry. At the time of writing this, a number of years since I closed that chapter of my life, it seems that more positive moves are being made to resuscitate SAPPEX. I am convinced that only a dedicated, full-time PRO/Secretary will bring SAPPEX back on track.

Of course, life carries on, and interesting publications came out in 2008, amongst which a feature on the Oudendijk Group in the Netherlands, a *Landbouweekblad* feature on Peter Dorrington, and an article on Boutique Fynbos showing superb plants raised by Hans Hettasch of Arnelia Nurseries in Hopefield. Honingklip Farm hosted the members of the Kogelberg Biosphere who found that '*Honingklip is an excellent example of what we strive for in the biosphere reserve; they use the natural resource, make use of sustainable technology and provide work for the local community.*'

It was gratifying to see that the ARC started writing articles in the *Farmers Weekly/Landbouweekblad* again.

International Developments

International Protea Association

After retiring from SAPPEX I thought I might as well go completely into retirement and gave notice that I would lay down my duties as Chairman of IPA at the San Diego meeting in April. I had already talked about the possibility earlier, but now I was sure that it was time. Having looked forward with anticipation to my last conference in the Chair, I was summarily told at our first Board Meeting by Hans Hettasch, acting on behalf of the new Executive Committee of SAPPEX, to relinquish my duty as Director representing South Africa. While digesting this sudden and unexpected move, I was requested, as a result of a SAPPEX Executive Committee decision, to stand down immediately as Chairman of IPA and to hand over the reins to Rua Petty of America who would be elected as next Chairman. I was so shocked by this development that I agreed, but later informed the Committee that I would be handling the Biennial General Meeting as outgoing Chairman. It was a sour note on which to end my Chairmanship of IPA.

My last input into the IPA was the publication of Volume 60 (May 2006) of *International Protea News*, the journal of the International Protea Association. It was unfortunately, the last one published.

Since then of course there was the highly successful IPA Conference at Stellenbosch with more than 200 delegates from nineteen countries. Rua Petty was Chairman, with Peter Dorrington as President of IPA and Hans Hettasch as Secretary. There were some nice innovative program changes, and it was a huge success. FloriCulture International (October 2008) published a two-page report, also showing the growing areas worldwide. It was quite a change to be able to attend as a delegate and not having to rush to different meetings, although I was invited to sit in on one or two to clarify some points on administrative affairs. One keeps learning at such events. Just when you think you know quite a lot, then new research brings new facts to life. Pity the poor scientists in years to come who have so much to learn!

There has been some very sporadic general information since then, but even the Biennial Conference for August 2010 to be held in conjunction with the ISHS Congress in Lisbon received hardly any publicity. It was just assumed you would somehow know where to find the information. Just prior to the Lisbon Conference, there was a change in Chairmanship, with Audrey Gerber having to take over at short notice from Rua Petty. Audrey already served as the Chair for the IPWG Working Group (Research). So she had a double whammy up to and including the conference in Lisbon. By late 2010 (where my saga ends) the new IPA web site was launched and it seemed that the IPA was on track once more.

Other International News

In FloraCulture International 2009, a small article appeared on *Protea cynaroides* patio plants that are grown by ten growers of the Flora Toscana cooperative in Italy with the plants being sold all over Europe. The initial trials of the small cynaroides variety were initiated by Proteaflora of Australia and Flora Toscana. Sea freight had become a regular way of transporting proteas. It was reported that in the Netherlands Oudendijk Imports alone imports many sea containers of protea, *Leucospermum* foliage, and *Hypericum* from South Africa, Portugal, and Equador. I am sure they are not the only ones to do so.

Niek Oudendijk of Oudendijk International, Netherlands,

In Australia, new quality standards were published for thirty-two wildflower products. Each specification includes a product description covering flowers, leaves, stem, and stage of opening, with clear photos illustrating stages of opening. A manual for post-harvest handing was also published. This includes product-handling advice for importers, wholesalers, and consumers.

And then came 2010 and the Iceland volcanic eruptions, which caused a huge disruption in flower movements around the world. Huge losses were reported with supermarkets in Europe having to cancel their orders. It was estimated that the crises cost the Kenya flower farmers around US$15.5 million per day. It could not have come at a worse time, just before Mother's day, which is traditionally the time to sell flowers

So at the end of 2010, I hope that sometime in the future, someone will write up the next chapters of the Protea Industry.

APPENDIX A:

Gert Brits : 'South African Proteas (Frank Batchelor starts protea cultivation in South Africa)

Extract from IPWG Newsletter No.1, August 1984.

The first person to plant proteas on a larger scale than merely for garden ornamentation was A.C. Buller, who grew *P. cynaroides* in the Banghoek Valley at Stellenbosch around 1910. In about 1920 another resident of Banghoek, Miss Kate C. Stanford, started the first commercial venture that sold protea plants to garden owners. From the late forties onwards W.J. Middelmann functioned as the largest supplier of Proteaceae seed.

These people were, however, mere forerunners of the great protea pioneer, Frank C. Batchelor, who developed the commercial cultivation of protea cut flowers in South Africa from the forties to the seventies. During this period Batchelor laid the foundation for commercial protea cultivation in South Africa, by means of the following:

- He started the first farming enterprise relying entirely on the production of protea cut flowers.
- He vigorously promoted the use of his cut flowers among the general public and succeeded in the building up a keen local demand for proteas.
- He became the first protea farmer to market cultivated proteas overseas.
- He set high standards for the picking quality of proteas.
- He was the first commercial protea breeder; for example he created the first deep red pincushions after about five generations of mass selection; the first protea cultivars registered internationally were five of his selections (1974).
- He was the first person who utilised natural hybrids professionally by using vegetative propagation — he created the first commercial vegetative planting in this way.

- He was founder member of SAWGRA (S.A. Wildflower Growers' Association — later SAPPEX).
- On his initiative the SAPPEX Batchelor Competition was introduced to encourage the breeding of proteas in die industry (he donated the floating trophy).
- He was the first producer to identify fully with production research; far-sighted, he put his farm and plant material at the disposal of the University of Stellenbosch and the Department of Agriculture for research purpose.

The publication of a textbook by Vogts in 1958 enabled the cultivation of proteas by a wider public, both within and outside the habitat.

APPENDIX B:

Some Aspects of the Problem of Preserving the Indigenous Flora of S.A.

Notes for a talk to the Cape Natural History Club by Capt. E.J. Scholtz on August 31, 1949.

Before starting on the actual subject of the talk it would be of interest to contrast events of today with those of a couple of hundred years ago in regard to the exploitation of our flora. Today, quite rightly, there is a ban on the export of our wild plants and flowers to other countries and we are debarred, again quite rightly, from collecting even for private use in South Africa; but in those good old days — or rather bad old days so far as the treatment of our flora is concerned — literally shiploads of our plants were collected and exported. No doubt science benefited materially in consequence, but doubtless also the permanent loss to South Africa was severe. We have proof of this when we read of AITON in 1975:

It may be noted, in testimony of the zeal and industry of this remarkable man, and despite the extent to which the Cape Colony has been traveled over by observant botanists since his time, and to which railway transit has rendered accessible the carroid habitats of these plants, there are still species figured by MASSON which have never been found since his time.

So you see that a century and a half ago we had through excessive collection lost, it seems for all time, a number of our rare species.

The following extracts from authentic botanical works offer proof of the extent to which plant collection took place in those old days:

The Dutch Settlement at the Cape of Good Hope could not have existed long without some striking examples of its wonderful flora being sent home in vessels touching on their return voyage from the East Indies. Holland was throughout the sixteenth seventeen and part of the eighteenth centuries at the head of the European horticulture.

We have the record of Paul Hermann, a botanist and physician of some note (in the eighteenth century) who on his voyage to Ceylon made the usual short stay at the Cape and, in the immediate neighbourhood of the Settlement gather together a collection, which Thunberg describes *Herbarium insigne*.

Of Auge it is written: In 1761 he was sent to accompany an expedition under Commandant Hopp to the Namaqua territory, and returned with a large harvest of plants. Eleven years afterwards he became the companion and guide of Thunberg & Masson in their collection excursions.

Of the great Thunberg himself, referred to as the father of 'Cape Botany', and who in 1783 succeeded the founder of the current system of botany, Lynnaeus, at the Upsala University, it is written:

He reached the Settlement in April, 1772. His sojourn at the Cape lasted for nearly three years, and of his indefatigable industry during that period we may judge by the fact that his Flora Capensis based on his own collections, enumerates no less than 3,100 species.

I come now to the subject matter of the talk. We all know what fine work has been done by Kirstenbosch, but it stands to reason that the concentration of the huge variety of plant life in one area, with its limited opportunities having regard to altitude, aspect, soil and climatic conditions, rainfall etc., cannot give the best results. A wider field than Kirstenbosch alone should be brought under experimentation.

For the propagation of succulents, we recently acquired the Karoo Garden and big things are expected of this development at Worcester. Succulents therefore do not enter into this discussion.

South Africa and the Cape Peninsula in particular are unique in their variety and beauty of wild-flower life. Was it not Thunberg himself who remarked that the Cape Peninsula was richer in wild-flower plant life than any other place of its size in the world. That being the case, it seems our thoughts should be directed to finding means whereby the activities of Kirstenbosch can be supported and supplemented in a way which would tend towards even better results than those that are being secured at present.

For this purpose, one calls to mind a government department whose functions and sphere of duty are very closely allied to the work carried out at Kirstenbosch. I refer to the Department of Forests. Apart from the actual duties of the department in the propagation of exotic trees and the protection of our valuable indigenous forests, there is the important factor — very important so far as our flora is concerned — to be borne in mind that all lands in South Africa that have not been alienated by the State for farming, town requirements, and other public purposes, are ipso facto forest reserves, even though there may not be a single timber tree growing on the reserve.

The following, therefore, is an attempt at setting out in what manner the activities of the forest department officials could be of immense value to the objectives for which Kirstenbosch is striving. The objection that at once comes to mind is that the department itself would be inclined to turn a deaf ear to projects that might materially — from a point of view of time — interfere with the normal function of its officials. But in the suggestions that follow, care has been taken to abstain from any proposal that would have this effect. The projects I have in mind can be classified under three heads, viz:

1. Foresters to study the wild flora of their particular districts;
2. Foresters to collect seed of species growing in their Districts;
3. Kirstenbosch to collate and make use of the information gleaned by foresters and the seeds collected by them.

When it is borne in mind that there a large number of foresters are employed in the Department of Forests, spread over the whole Union, each forester at present taking a keen interest in his particular area from a tree-growing and forest-preservation point of view, and under this scheme becoming a specialist as regards the natural flora of his area, the possibilities of advancement in the field under review will be apparent.

1. Foresters to study the flora of their particular districts:

The first duty of the forester would be to compile a list as complete as possible of all the species growing naturally in his own district. The forester in the course of his normal duties has to patrol the area under his control, so that a compilation of the flora of his district would not impose upon his duties occupying much of his time. It is obvious that it would be most valuable to have a record of the habitat of each of the species of his area, many of which doubtless are being grown at Kirstenbosch at present. As was stated, many species collected by Masson 150 years ago have never been found since. We cannot assume that these have become extinct by reason of veld fires or excessive collection by those early assiduous botanists. It surely is not unreasonable to expect that some, at any rate, of these species which have been lost to us over this long period of years will come to light as the result of the study by the forester of the flora of his area. It follows also that such a study will result in new species being discovered of which we know nothing today.

The second duty of the forester would be to make notes of aspects like soil, altitude, climatic conditions, rainfall etc., of the localities in which the different species grow for the purpose of advising Kirstenbosch and thus facilitating the work of their cultivation in our national gardens, helping Kirstenbosch to advise members of the Botanical Society of the proper conditions for producing good results, and ensuring better results in the propagation of these species when definite data are available following the study of them under natural conditions.

The third duty of the forester would be to keep a watchful eye upon the localities where they grow, and to make notes as to whether the areas occupied by them are extending or diminishing; if the latter is true the forester has to endeavor to ascertain the reason for their diminution; are the areas diminishing due to fire, theft, destruction by cattle, unfavorable seasons, etc.? It is not only important to know whether a particular species is thriving or diminishing, and if possible the reasons for its failure, but the opportunity in such a study would be afforded of attempting its growth in the Forest areas with more favorable conditions and minus the condition causing its retardation in its natural habitat. Here may be quoted the case of the Blushing Bride (*Serruria florida*), which it was felt some years back was in danger of possible extinction owing to devastating fires in the natural habitat of this species in the French Hoek District. Although, as events proved, there was no danger of this happening, it might well have occurred in respect of some other species.

2. **Foresters to collect seed of species growing in their districts:**

As previously stated, the forester must in any case patrol his district and in doing so it would be a simple matter for him in season to collect seeds of his natural flora and to sow these when sowing his exotic tree seeds. Thereafter, small nurseries might be established in the neighbourhood of his quarters, and a few of the transplants might even be planted in suitable situations. The opportunity would thus enable to make a comparison between man-made-plantings and plants that have grown on its own accord naturally in his district.

Seeds collected by the forester in his area should also be sent to Kirstenbosch for trying out in our national gardens. Flowers and foliage of such seeds should go to Kirstenbosch so that the species may in the first place be determined there. It would be of interest to compare the results obtained at Kirstenbosch with their natural growth in the area of their origin and the plantings-out in such areas.

Seeds should be sent to other foresters, in exchange, for the purpose of further testing out.

As the result of these various activities on the growing of seeds collected by the forester, valuable data would be made available to the scientific centre at Kirstenbosch, and inter alia Kirstenbosch would be placed in a much better position than today to advise members of the public generally of the species which, after this wide range of experience, would be best suited for growing in private gardens in different parts of South Africa.

3 Periodical reports to Kirstenbosch:

The information so gleaned by foresters in the close study of their flora in their respective areas and the results obtained by them in such areas of propagation there of the seeds they have collected should be reported, say annually by them, or through their Senior Officer, to be made use of for scientific and other purposes as deemed desirable by the Director.

I make bold to say that foresters will be none the worse Forest Department Officials for the interest they are called upon to take in studying and experimenting with, within reason, our wild flora, nor would the Forest Department be in any way the loser by the time occupied by the foresters in such activities. On the other hand, South Africa would be vastly richer in the important subject of the better preservation and the wider use of our glorious wild flora.

The results kept by Kirstenbosch as the result of the activities sketched above could be made use of for the following purposes:

a) Keeping a check on species where there is danger of their becoming extinct;
b) Introducing such species into other areas where they are likely to succeed, and where danger of extinction does not exist;
c) Making use of the practical experience of hundreds of trained foresters, located in various Districts throughout South Africa, in our National Botanic Gardens.
d) Encouraging the growing of our wild flora by members of the Botanical Society and the public.

I would now like to enlarge somewhat upon the last-mentioned uses, which Kirstenbosch could make of for the proposals put forward, i.e., encouraging the growing of flora by members and by the public. I have always felt that the growing of our veld flowers in our gardens is one of the most important factors for the preservation of our flora. Where this is done, a greater respect is engendered and the inclination to collect flowers indiscriminately in the veld is lessened. Little is known of the charm of our veld flowers in the garden and much could be done to encourage their propagation in home gardens. The practice would be of distinct advantage to the coloured people who make a livelihood from the sale of such flowers, if they were taught what species to grow and how to grow them. We rightly take strong measures to protect our flora by punishing those who collect and sell our protected species; personally, I should like to see more encouragement and help given to the coloured folk to grow their own. Our aim should tend more to educative rather than punitive measures. Then, also, by growing these flowers ourselves, it surely is not unreasonable to anticipate that some at any rate of the species which in the course of time would have disappeared from our ken would be saved.

But perhaps the most important reason why we should grow our wild flowers in our own gardens is the sheer delight such a hobby affords to those of us whose favourite pastime is gardening.

Hobby of Wild Flower Gardening.

I feel sure the hobby of gardening in one form or another is one which is indulged in by a large percentage of the members of our Natural History Club than any other hobby. And I would like to encourage those members to switch over to wild flower gardening, because if you find ordinary gardening enjoyable you would find specialising in our native flowers a most fascinating hobby. So little is known of it, that our efforts are largely pioneering work; we are still largely in the experimental stages. And all this simply adds to its charm.

At the start I want to emphasise that there is only one way of making a success of the pastime, and that is growing your subjects from seed. Nature protects herself; if you set out with the object of obtaining your plants from the veld — rooting out what appealed to you and transferring them to your garden, the attempt would end in utter failure. Some have the idea that in wild flowers grown in our gardens we do not get such splashes of brilliant colouring as are provided by domesticated species. This is quite wrong. There surely could not be a finer display of colour than is to be seen at Kirstenbosch among the *Leucospermums* (pincushions).

My own experience as the years roll by and the opportunities for spending happy hours on long tramps in the veld and on the mountain amongst the unique flora of the Cape — as these opportunities become fewer, there is only one thing for it and that is to bring the veld and mountain into your garden.

Do not let what appear to be unsuitable conditions for growing veld subjects deter you from taking a shot at it because with experience you will find some at any rate responding to your efforts. I started fourteen years ago and somehow in some respects I had better results then, than now — beginner's luck no doubt. But when one compares the uphill work it is establishing a wild flower garden at a place like Muizenberg with its sandy soil conditions needing constant waterings in summer, low rainfall, and high winds with conditions for example enjoyed by Mr. Batchelor on the lower mountain ranges of Stellenbosch — well you cannot help being very envious. His conditions are extremely favourable — he never for instance waters his proteas or other shrubs after planting out into the open. Despite this, his percentage of losses is negligible, and the growth of his plants infinitely finer.

For results — blooms from your favourite subjects if these be members of the proteaceae family — one has to exercise much patience. Most proteas and *Leucospermums* produce their first blooms four years after sowing and in some cases one must wait double that time. But you are amply repaid for this long wait once they start blooming. A *Leucospermum* nutans for example, while giving you only half a dozen or so blooms in the first year of flowering will annually produce flush of several hundred blooms a few years later, of exquisite colour and shape, and the flowering period continuing for several months. A feature of the flowering of the *Leucospermum* genus, and contrary to what applies in proteas, is that as each bloom finishes its flowering period, it drops from the tree and its place is taken by another flower equally as beautiful and leaving no trace whatever of where the earlier flower had appeared.

Disappointing set-backs you will have, some defying anything remedial. *Leucospermums* baffle one — some are hardy from the time of germination and give you very little trouble — amongst these I count *Leucospermum incisum,* which as you know hails from Wolseley. The **Ls. tottum** from Bains Kloof, Villiersdorp and those parts, has a habit of just petering out after its first show of blooms. **Ls nutans** — Houwhoek and Caledon's lovely pincushion — contracts a fungus disease after seven or eight years, and while it is possible to keep it in check to some extent by spraying, the plant is not worth retaining once the disease becomes apparent.

Sequence of events in growing of Veld Flora

The site of your *seed beds* is an important factor. Choose a spot comparatively level where in heavy rains your seeds would have drainage but not be washed away, and where your bed would have sun in the morning and natural shade in the afternoon.

Soil. My normal soil conditions: Very sandy. For humus, I have pits into which all the weeds go. After a year or two the product is used not only for the seed beds (after sifting) but the unsifted material is used for filling the holes to receive the transplants from the seed beds.

Sowing. Seed is available from Kirstenbosch from March onwards. Towards the end of March and for a few months afterwards is a good time for sowing. As we are by no means sure as to the actual depth at which seeds should be sown, it is a good plan after preparing your seed bed to rake your seeds in so that varying depths occur. In all the many steps to be taken, follow nature as closely as possible, and for this reason it is useful to take careful note of conditions where the plants have their natural habitat. A small point but a useful one is to observe how a protea seed with its feathery surround descends to the ground from the flower. Plant it with the point down in the same way — with some hair above ground — not too deep. Very little handling of the fistful of seed from year-old flowers will soon teach you by the feel which seeds are fertile. The fat, stocky little chaps produce the goods. Remember that you must wait a year for seeds from proteas, from the time flower is at its best; whereas with *Leucospermums* (pincushions), your seeds are deposited from the flower very shortly (a week or two) after the flowers fades.

Birds, field mice, and ants one must contend with when you do your sowing. The former two can be kept away from your seeds by taking small gauge wire (canary cage gauge) and you make cones for standing on your seeds so that birds and mice cannot enter. For ants to be kept at a distance DDT is useful. How much water to apply to your seed beds depends largely upon weather conditions, your type of soil, and the type of seed.

Planting out from seed beds. Planting out in the situation chosen for your plants must be done the same season as that in which the sowing is made, and the planting out should be done as early in the year as possible so that in their permanent site the young plants may receive the maximum benefit of the season's rains. Those of you who have grown sweet peas will know the exceptional length of root-growth to the growth of the green stem above ground. The same thing applies to proteas. As in their natural conditions they must at the end of the winter when rains cease, do without rain for seven or eight months, nature provides a good length of root growth to lower depths where moisture continues. With sandy conditions, such as we have at Muizenberg where I live, plants must be watered constantly during summer and particularly if it is a specially dry one — once a week and even more often. It is a good thing to put some of your seedlings into earthenware pots and let them develop in them for a year until the next planting season, but personally I think it does a tap-rooted plant no good whatever to cut back the tap root for pricking out into these pots. Nature did not provide a taproot for nothing — to get to deeper and reach more moist levels in the non-rainy period.

If the taproot is removed and lateral roots encouraged the exposure to the drier surface conditions of the soil must be detrimental. Before planting out from the seed bed, make a careful study of the size and habit of your plant to ensure your selecting a suitable 'possey' (as the Australians have it) for your plant. How often one kicks oneself for planting too close to a path where the plant lacks elbow room and again where a dainty flower is placed further back and does not show to advantage when one saunters up the garden path. A very good illustration of what I have in mind is planting protea rosacea (our skaamblom) anywhere else but right beside your path and on an elevated situation so that you may feast upon the beauty of this flower when you bend down and look through it to the setting sun. The finest colours in stained glass windows under the most favourable conditions give but a poor effect in comparison.

Early growth and preparation of holes to receive transplants. Where poor soil — sandy — has to be coped with, it is important that holes be made, say two feet deep by two feet wide, in which prepared soil is placed before planting out; soil from your humus holes mixed with say a good type of top-dressing soil. After planting, covering the soil with a good layer of straw to keep the soil moist is a distinct advantage. One would imagine that a so-called hardy veld plant would need but little water. My experience is that, if anything, it needs even more than the commoner garden plants of everyday use. But here again it all depends upon your soil conditions — the nature of your soil. If I do not water my plants right through the summer, the growing of veld plants would be a total failure. At Devon Valley in Stellenbosch, Mr. Batchelor does not water even at the time of planting out and his losses are, he tells me, not as much as 5%. Shading of your transplants is most important and some types need shading — in summer — for years. In the absence of shade they generally have a very short life; and one notices that where shrubs grow alongside, or even longish grass, and shade is thus afforded, the removal (unless very gradually done) of such protection frequently means death of the plant immediately after removal.

General:

Patience has to be exercised in awaiting the flowering period. Most Proteaceae take four years from the time of sowing before the first blooms appear. But from that time onwards you are amply repaid your long wait. The life of proteas is said to be from 9-12 years. Many species develop abnormally slowly, but these generally after reaching a certain stage just spurt ahead later, and have a longer life; the same thing occurs in the slow growing stinkwood with 400 years' growth to maturity, and producing the most exquisite of all our timbers, in comparison with the soft wood pine timbers with only a forty years' growth before felling.

Annuals.

It is very well worth while to grow amongst your bigger shrubs, patches of annuals of the Namaqualand daisy type: heliophyla, sutera, arctotis, dimorphotheca etc. Obtain a few packets of seeds (if you are a member of Kirstenbosch you will find their seed excellent) and sow them into your seed beds, and on a suitable rainy day, pick out and plant all over amongst your shrubs. If the planting out is done at the right time and if weather conditions are favourable you will be well repaid that season; but the charm of the flowers is that they propagate themselves very readily, so that the next year and years to follow a gorgeous display of colour is afforded, without your having to do to bring this about, expect for scattering stalks when the seed attached to them are about to be deposited on the ground. Watering of these is not necessary, because they spring up with winter rains and come into flower in spring. If you should have a couple of bad seasons on account of poor rainfall repeat the process, but in any case it is desirable to obtain fresh seed and sow say every five years or so.

And here I think it would not be inappropriate to pay a tribute to Kirstenbosch. We, living in the glorious Cape of ours, are singularly fortunate in the lovely natural floral surroundings we have for continuous enjoyment. But little appreciation I am afraid do we show for it. And then to cap it, Kirstenbosch is there to give us a fine selection of seeds of beautiful veld plants as one could possibly desire, and at a cost which bears no comparison with the value of those seeds. There is no other centre in South Africa where anything similar is done, and I make bold to say that nowhere in the world does a garden-lover and moreover one specially interested in his native flora enjoy a similar unique privilege. We should have a membership of our Botanical Society at least five times what it is today.

In conclusion I may quote this sentiment, with which I know every one of us will be in full accord:

The reward, which constant association with nature's gifts and beauties grants to those that love her, is far beyond all other earthly remunerations.

Muizenberg, July 1949.

APPENDIX C:

The story of SAPPEX

(South African Protea Producers and Exporter Association)

As published in Veld and Flora 29 Sept 1984

Sappex was born in 1965 out of the need to deal with some controversial matters such as: Should protea seeds exports be prohibited to 'protect' a newly developing flower industry? Should the State be allowed to sell protea plants in competition with recently emerged protea plant nurseries whose prices were allegedly 'too high'?

The history of local development of proteas in horticulture and floriculture is dealt with elsewhere in this issue. Suffice it to say that until Kirstenbosch started in 1913 nobody in this country took much interest. Another 30-40 years were needed until the general public recognised the value of our native flora and only then people started wanting to buy some plants to put into their gardens and to use proteas to any extent in floristry. By the early sixties overseas demand for protea flowers added to this interest and commercial protea growing started in earnest.

A couple of small plant nurseries had commenced and encouraged the demand when a State organisation started to climb on the band wagon. Other well-meaning but poorly informed people during 1964 thought that if protea exports would develop, a monopoly for South Africa could be created by prohibiting seed exports to potentially competing countries, a measure which would have been impossible to administer.

Firmly believing in private enterprise, W.J. Middelmann whose wife at that stage conducted a protea nursery and seed business got together with pioneer grower Frank Batchelor, who sold flowers, plants and also had exported seeds, to discuss the proposed measures. We strongly felt that only an Association could put the case and, to gain support and secretarial help, contacted Alan Starke, then Chairman of Starke-Ayres Ltd, about forming an organisation.

Thus, on 18 August 1965 and with the support of Dr. Ruben Nel, then the Director of the Fruit and Fruit Research Institute in Stellenbosch (FFTRI), Walter Middelmann as Convenor called a Meeting of interested parties and the South African Wildflower Growers Association (SAWGRA) was formed, affiliated to the S.A. Nursery men's Association which offered secretarial help. Soon there were 80 members. Prof. Brian Rycroft, Dr. Marie Vogts, the late C. Meiring of Caledon Gardens and the late Dr. Ruben Nel were made honorary life members.

At that stage plant nurserymen were strongly represented; flower exports had not yet really taken off. The battles regarding plant sales and seed export prohibition were soon won. Other activities included railage costs, marketing and in particular the application of the Cape Nature Conservation Ordinance. SAWGRA arranged regular Meetings, lectures, films and excursions such as to the then pioneer experimental plantings of *Protea cynaroides* at Oudebos (Marie Vogts), Protea Heights (F.C. Batchelor), Highlands Forest Station (Forest Dept. seed supplies) etc. Even a trip to Johannesburg to visit the Multiflora flower market and the glasshouse flower industry there was arranged. Already in 1969 Harry Wood, another pioneer protea grower and on the Committee since its inception suggested a competition for 'self-owned and bred hybrids', out of which the Batchelor Prize developed with the intention to encourage vegetative propagation. Liaison with the government Protea Research Unit, at that time based at the FFTRI under Dr. Marie Vogts became important and the FFTRI to this day hosts our Meetings. In the early 1970s SAWGRA was invited to the Nature Conservation Advisory Committee in connection with the revision of the Ordinance though this membership was later, regretfully, terminated.

During the early days, undoubtedly, nurserymen played a leading role and SANA's support and secretarial services, use of its Journal, were important. Gradually, however, the industry's activities became more and more orientated towards cut-flower exports. Such protea exports amounted to 144 tons in 1971 but quadrupled by 1974. By 1982 it was 1925 tons. Representations in connection with airfreight, inspection standards etc. led to recognition of the Association's role and it was actually government which suggested a name change to South African Protea Producers and Exporters Association (SAPPEX). This was accepted just over ten years ago at the 6 August 1974 AGM, with a new Constitution. The SANA affiliation then fell away.

During that period a Newsletter was started which, under the editorship of Harry Wood (1974 until resignation 1980), became the leading publication of its kind, worldwide. There was regular collaboration with the Dept. of Nature Conservation, especially prior to the promulgation, early 1975, of the new Ordinance. One of SAPPEX's aims in the Constitution reads: 'To uphold the cause of Nature Conservation, where applicable, and especially in connection with optimal utilisation of floral veld as a sustained natural resource'. It should be mentioned that SAPPEX is also represented at the Flora Conservation Committee of the Botanical Society of S.A.

With the development of protea plantations, the need for management techniques, for achieving better quality standards, freedom from pests and diseases, post-harvest treatment methods, research liaison became ever more important and a Committee, since 1971, under Mrs. Anne Gray has closely worked with Gert Brits of the Research Unit at Riviersonderend, Dr. Gerard Jacobs of the University of Stellenbosch, Dr. Sharon von Broembsen (diseases) and others. Field days have been held and enabled those in the industry to meet each other while attending talks and demonstrations on progress made. With effect from 1984 SAPPEX has now also been officially appointed by the Horticultural Research Institute to handle the distribution of their releases of cuttings of approved variants.

By 1976 Walter Midddelmann resigned as Chairman opening the field to a younger generation. He has been elected as Hon. Life President of the Association. Stefaan Joubert took over for two years and during that period very generous support of Indo-Atlantic Air Cargo on the secretarial side helped SAPPEX to get on a better financial footing. Barrie Gibson, the present Chairman, took over August 1978 and when three years later the outside support ended and Maryke Middelmann took over as Secretary. The introduction of a Levy System based on cartons used in the industry finally put SAPPEX on a sound financial footing. SAPPEX has since been able to bring out some promotional material such as brochures on commercial varieties and large posters.

In particular SAPPEX has been able to create a Research Fund and to allocate research grants on a regular basis, generously supported through the S.A. Nature Foundation, to scientific institutions. It has also made possible SAPPEX participation in the person of its Chairman, Barrie Gibson, at the second International Protea Conference 1983 on Maui, Hawaii and Seattle, U.S.A., and, again with the support of Nature Foundation, the sending of Gert Brits to the same Conference. Further financial support, also with Nature Foundation, has been forthcoming in connection with bringing overseas scientists to the Inaugural Meeting of the Protea Working Group of the ISHS through which, in turn, the joint IPA 1985 Conference and ISHS Protea Working Group Symposium have become possible.

International connections originally established by Walter Middelmann not only consist of SAPPEX membership of Union Fleurs which represents the European flower trade but in particular collaboration with the International Protea Association (IPA). The enthusiasm of Dr. Philip Parvin, Hawaii and Peter Mathews, Australia, for the protea cause, expressed in several visits to this country, has led to the formation of this international organisation and through SAPPEX will act as host country for the Conference to be held at the Cape in August/September 1985, with SAPPEX Chairman Barrie Gibson acting as IPA President.

This issue of Veld & Flora in many ways covers aspects to be dealt with at that event, unique because of its accent on excursions to see where the proteas actually come from (habitat) and to research institutions, practical growers as well as marketing organisations.

Today SAPPEX, twenty years old by the time of the Conference, with well over 200 members, 38 of them from other countries, stands as the recognised mouthpiece of a developing new industry regarded with worldwide interest.

W.J. Middelmann

APPENDIX D:

The birth of the Dried Flower Industry in South Africa

by Walter Middelmann 23.01.2003

Gerald McCann, former (long ago) Forester, Lebanon Reserve, who did a lot of walks with Ruth over Forestry (Govt.) Nature Reserve Ground looking for Proteaceae and other Fynbos plants for identification now has written a book about his work at the time. Ruth, Honingklip and the Middelmanns play quite a part in this and he has asked me for some details.

I used the opportunity to try and find dates regarding Ruth's dried flowers inspiration, the real start of our enterprise. It has always been a subject to tell friends in general terms when asked how it came about. So I went through my Honingklip Diaries, diaries written up in my lousy handwriting, usually late at night after hard work, in a very rough fashion, yet as far as I can see kept complete right from the start and still continued with a short note after each visit, generally now only twice a year and for social reasons, Christmas and so on. Yet I at least made a note of something new or special I saw or felt or have been told.

So here we go; Ruth had, in the early sixties already collected some 'cone branches', i.e., Leucadendron foliage with seed cones, hung up, turned brown and called by us '*Brown Stuff*'. It seems she managed to sell some through her fresh flower (Protea) florists and there are probably records in old invoice books somewhere. Certainly Everlastings were also collected and sold locally on a small scale. All this was suitable to give the workers' families extra income with rewards based on certain quantities for various articles, 'Stukwerk' or 'piecework', paid out on weekends when we got to Honingklip, then hung up in our first store building, of own design, built by me personally with assistance from the labourers just behind the house.

Gradually quantities became large and overseas markets were developed for shipping in cartons, the very first ones to London by sea parcel post.

While originally it was branches that had already dried in nature we then started cutting well formed ones to be dried on a larger scale, the more so since we later (1964) had visits from Eunice Curtis, Seattle, and the Wallace Stevens from Wanganui, New Zealand, who had suggested deliberate selecting on certain standards, drying, packaging as seen for other floral products in other countries. Notes on these events need to be looked up.

I find that 15.10.1963 we had a visit from Rolf Jülicher, Johannesburg carnation grower and exporter, suggested by Dr. Fred Anderssen, prominent Agri-Scientist with whom we were in touch about Proteaceae seed export problems. Rolf wanted to see a Protea Farm with the idea of proper floriculture, propagation by cuttings, crossings etc., which of course we, non-resident at the farm and without experience could not envisage. However, he put us in touch with Kervan in New York, flower importers, suggesting our Everlastings to them. This led to myself visiting there and through them to others in U.S.A., to start with for Everlastings but then on a wider 'brown' foliage/flowers field.

A major chance event came 21 January 1964 after a big fire caused through the Forest Dept. burning belts and the fire jumping over, burning a large section of our mountain. Ruth went to collect fallen Protea seed heads for her sorting out the seed contents, also some 'grasses' — Restios — which we had bundled and which got sold to florists. A London man producing wrought iron flower stands came to our door in Newlands one day asking whether we could send him parcels of material he had seen at a Sea Point flower shop. We sent such and the products were seen in London by De Mooy, a Dutch florist wholesaler. Thus we soon found markets for the 'brown stuff' at De Mooy, Netherlands in, to us at the time, seemingly huge quantities which knowledge and demand then spread out rapidly to Meschi, Italy, to Japan, Germany etc., and so also the assortment got widened. Old invoices are at Honingklip, where Maryke started and continues building up a historical library of Honingklipeana.

Diary entries 14/15.03.64 and again 28.03.64 refer to this, now also mentioning '*Stars*' for the first time, as 'collected in big fire areas'. There is mention of 'still trying to settle the packing problem for all the masses of dry stuff', also 'cartons hard to get'.

And then under that same date, *28 March 1964* the entry: 'found beautiful golden leafed burnt compacta flowers, also sugarbush stars collected on Monday'.

I feel that THIS was RUTH's REAL INSPIRATION DATE for the further large scale development of our activities, Easter Weekend 1964.

23.01.2003

APPENDIX E:

South African Protea Producers and Exporters Association
Discussion Day
22.09.1976
Objectives of the Cape Ordinance on Nature Conservation

A background sketch of aims and objectives — DR. HEY.

First of all, Mr. Chairman, I want to make it clear that our task is Nature Conservation and we interpret Nature Conservation in the sense of the wise management and wise usage of our resources of flora and fauna.

As such we are willing to co-operate with any Organisation where we have common cause, but I don't think you can ask us to undertake things or introduce elements into our Ordinance which will allow loopholes for malpractice.

I want, first of all, to assure you that the door to my office and that of my Officials, are always open to you and your Members, either individually or as a deputation of your Association. We are always willing to discuss matters with you in order to try and establish where we have common ground. I would, therefore, briefly like to give you the background of this entire situation so that we can go forward constructively, both as far as Nature Conservation as well as your Association is concerned.

The first thing I want to say is that we support very, very strongly, the concept of cultivating wild flowers. Mr. Chairman, I would just like to say that, as far back as 1962 (when you were still in school), we were working on Conservation. At that time we started the concept — *GROW YOUR OWN — KWEEK U EIE!* At the Van Riebeeck Festival in 1962, when Nature Conservation first started, we had a stand in the Pavilion based on this concept — Grow your Own — which we managed in close collaboration with Kirstenbosch. The idea was that, by promoting this concept, we would reduce the pressure on our natural resources.

So you can see that, right from the outset, we were with you, and when this idea started to develop we decided to have another look at the Ordinance.

Initially the Ordinance was designed as a purely protective measure, but gradually we have tried to interpret this Ordinance in the spirit of wise management and wise usage. We realise, of course, that if we really want to promote this concept, we would have to adjust our Ordinance accordingly and, therefore, provisions are made in the Ordinance, to ensure that this is possible.

We did this by providing, in the Ordinance, for the registration of Nurseries and Registering Stations from which Wild Flowers could be sold.

Initially the Ordinance was designed as a purely protective measure, but gradually we have tried to interpret this Ordinance in the spirit of wise management and wise usage. We realise, of course, that if we really want to promote this concept, we would have to adjust our Ordinance accordingly and, therefore, provisions were made in the Ordinance, to ensure that this is possible.

We did this by providing, in the Ordinance, for the registration of Nurseries and Registering Stations from which Wild Flowers could be sold.

The idea, and I would like to make it very clear to this meeting, has a two-fold objective. In the first place we are concerned, basically, with flora and fauna Conservation but, at the same time, it is also, to a certain extent, a protection and support to all Bona-fide Growers.

Met ander woorde, Mnr. Die Voorsitter, dit is 'n verskansel onder ons Wetgewing, vir diegene wat werklik ernstig is met die kweek van ons inheemse blomme.

The next stage came into being when people came to us saying that, as they already have ample supplies of Proteas growing on their farms, they do not need to cultivate them. Why should they then register to grow if they have whole hillsides covered with Proteas, ready to be picked and sold?

This was indeed a very reasonable request and we then decided to interpret the definition of 'cultivate' in a much wider sense so that it would include, as far as the law was concerned, the fencing in of areas which were protected from burning; the cutting with secuteurs; pruning; and all things concerned with the control of alien vegetation.

The next development, I am trying to show you that we have, indeed, co-operated right down the line and, quite naturally, I am pleading for my Department, arose when certain people approached me with the request: 'I do not have enough proteas growing on my farm for my business. I must, therefore, be able to hire property on which I can cut or get cutting rights. There are farmers who are not interested in proteas and would rather burn or plow them. If I can, therefore, hire the ground or get cutting rights, these proteas will be preserved and my business can grow.'

A concession was then made for the hire of ground and cutting rights.

In this vein, over the years, there have been repeated cases, I can go on Ad Infinitum, where we have had consultations with individual members or met with deputations from the growers, to try and help as much as we could. Names that come to mind here are

> Mr. Walter Middelmann who has been to my office many, many times;
> Mr. Harry Wood;
> you, yourself Mr. Chairman;
> Mr. Brink
> and Mr. Smal.

The next concession made, came as a result of our starting nurseries in order to promote the concept of 'Grow your own' and we needed to cultivate wild flowers which could be used for developing our Nature Reserves; re-stocking depleted areas and various other purposes, and then selling off our surplus to private individuals.

A protest from the protea growers was lodged with the administration saying 'They're taking the bread from our mouths'. We then again conceded by stopping our sales to the public and now only supply Provincial Institutions and Employees.

Finally, we come to the present ordinance which has been under very very heavy fire — not only from protea growers, but from all kinds of groups of people. It seems that the public who are not aware of the procedures adopted in the compilation of an ordinance, are under the impression that, we — of the Department, suddenly decide in a very bureaucratic manner, that the time has come to change the ordinance, so we sit down and write what we think would be good for the public and then say 'Here is the ordinance — take it or leave it'.

Mr. Chairman, this is indeed very far from the truth.

Do you know that this — all these things, start from a little black book which is kept in our office?

In this book, the Administrative Officers working with the ordinance, write down all the difficulties — all the items which are causing hitches of problems.

When we see that this book is getting a bit full, we realise that the time has come to take another look at the ordinance. It is only then that we sit down and start preparing a draft. Once we have a rough draft, we start consulting. Consultation on the present years extended over three years and you can have no idea of the number of consultations and meetings which were held, before this Ordinance was finally accepted.

As I have said before, the main function of the ordinance is to protect our natural resources of flora and fauna, but at the same time it also serves to protect and assist wild flower growers.

We believe that the existing ordinance provides these people with a certain amount of protection;

in other words, it allows the man who genuinely grows; genuinely picks and genuinely cultivates in terms of the definitions of the Ordinance, the opportunity of marketing his product.

Now, if we were to remove all restrictions, Mr. Chairman, then you must tell us how we are to protect the wild flowers that we are concerned about which could be obliterated — rare and endangered species that could be picked and sold. How are we going to know whether any flowers that are being sold and marketed — if there is no restriction — are being legitimately sold and marketed? This, Mr. Chairman, is our main and biggest problem.

What we are asking, Mr. Chairman is that (we have not received your Memorandum to date) you give us constructive ideas which we can work on and study to see whether any changes can be made. Providing amendments can be accepted what would be the consequential amendments. These things inevitably have repercussions — sometimes a very small change makes nonsense of the entire document.

We have to call in our Legal Adviser to tell us whether it is possible to bring about changes to the ordinance. All this we can do, but it is not up to us to make the final decision.

The matter gets referred to the Executive Committee, then to the Provincial Council and, by the time it has passed through all the various channels it may be a very different document than what we originally proposed.

Therefore, it was felt that no real purpose would be served in our coming here today to discuss possible changes to the ordinance as we are awaiting your Memorandum and need the time to make a careful study of this document.

This, I think, was the crux of the whole misunderstanding. It is not that we were unwilling to attend this discussion, but merely that to attend and not to be able to contribute any worthwhile information or answer questions about possible changes as suggested in your Memorandum but about which we are still unaware, would have been quite pointless. Once we have had time to study the Memorandum, then we can meet and discuss this matter fully.

We welcome your discussions — we welcome constructive suggestions about how we can promote the work you are doing, but at the same time, we must make sure that we do not fail in our duty towards the conservation of our fauna and flora.

APPENDIX F:

Survey Wildflower Production, 1981

It seems to be a twisted plea of a 'reliable source' to environmental bodies in Europe to promote the economic boycott of S.A. wildflowers (SAPPEX Newsletter March 1981). Firstly the motives of the writer seem twisted — pleading for ill will and destruction in the name of conservation. Secondly his facts seem twisted — for instance, the writer projects the shrinkage of the Fynbos by 61 per cent as if it has been an unabated process over the past 300 years. A fairer perspective, however, is to understand this as a historic process which has now slowed down greatly because of the remaining area consists of agriculturally and industrially uneconomical mountainous terrain. Also because large areas of mountain Fynbos managed by the Department of Water Affairs, Forestry and Environmental Conservation as water catchment areas enjoy a protected status. The writer is therefore dishonest in implying this process as continuing at the same rate all the time.

Faced by this spitefulness and by much emotion aroused by the fate of the Fynbos nowadays, your Association my find it worthwhile to press for the investigation of the relationship of the wildflower industry to the Fynbos. This has become an issue which could be resolved only to the benefit of all bodies involved: the sooner an authoritative study of the issue could be made, the sooner speculation and perhaps slander, may be checked.

The issue is divisible mainly into, firstly, a botanical study of the effects of the industry on the Fynbos, and secondly an agro-economical study of intensive wildflower production. The first part takes care mainly of the veld picking branch of the industry and the second mainly of the branch of scientific growing of wildflowers in flower orchards or plantations.

The planning of the botanical part of the investigation has already progressed far. It may be hoped that a proposed botanical survey may be launched soon. However, the agro-economic part has been lagging behind. It is this aspect which, I think, has now become the particular concern of the industry.

A proposed agro-economic study will not only help to neutralise ignorance and spite. The positive motives for such a study may definitely outweigh the one above as follows.

Intensive wildflower production as an essential element in a comprehensive Fynbos conservation policy.

The idea of encouraging the intensive production of wildflowers as a viable alternative to Veld picking was first publicly expressed in 1920 by the then newly established Wildflower Protection Society. Their ideal of creating 'a planting brigade rather than a plucking brigade' has been well quoted . . . The value of this idea is more evident today with the pressures of a vigorous wildflower export industry exerted on the Fynbos. In addition the idea has gained material thrust with the development, especially during the last two decades, of a small but ambitious 'planting' industry. This industry of wildflower cultivation, especially of proteaceae, in scientifically managed orchards presently contributes an estimated 10-15 per cent to the total of the wildflower export industry. It may, if optimally developed, eventually replace the veld picking of certain species to a large extent. The basis of this development is of course that planting should become more rewarding than plucking to the producer.

The value of intensive wildflower production is that it may have a long-term protective effect on a large number of Fynbos-species by means of negligence of the Fynbos. A case in point is the cultivation history of *Serruria florida* R. Br., endemic to the Franschhoek-area. Once considered extinct it was rediscovered near Franschoek, which exposed the species to potential overexploitation by veld picking. Due mainly to its purposeful introduction to cultivation by the Kirstenbosch Botanic Gardens, the 'blushing bride' protea then gradually became commonly available. Presently *S. florida* is comparatively neglected and safe in its habitat. Many times its available volume in the Fynbos is now produced in flower orchards.

The over stressing of negative measures to conserve the Fynbos may become counterproductive. It is important to also emphasise positive approaches, for example the creation of protectionism towards the

Fynbos and, especially, the promotion of intensive (orchard) wildflower production. The value of financial reward as a motive in the latter measure can hardly be overstressed.

Unfortunately, it is often not too obvious to the would-be intensive producer that 'planting' is more lucrative than 'plucking'. The temptation if the almost costless use of the veld often outweighs the long-term rewards of orchard cultivation. The desired intensive development therefore depends much on proper strategies to:

1. develop effective horticultural techniques of wildflower cultivation
2. breed for and provide superior marketable material to the industry and
3. to educate the producer community and promote orchard cultivation by means of an organised campaign.

This change through persuasion can come about only if the present intensive production industry is adequately understood. Too little is now known about its size, scope, structure and problems to influence its development meaningfully. The acquisition of a basic body of information may therefore be regarded as a prerequisite to direct the development of the intensive wildflower production industry purposefully.

Once successfully launched through its initial 'teething' stage, the future swing to orchard cultivation may accelerate at an increasing rate, due to the mounting competition enforced on the veld picked product by the superior cultivated product.

The promotion of a viable intensive wildflower cultivation industry should therefore be an essential element of any comprehensive Fynbos conservation policy. However to attain this goal, elementary information about the existing system of wildflower cultivation in orchards is needed. This may be acquired (preferably) by means of an agro-economic survey.

Justified use of Fynbos by the farming community for wildflower production.

The prerogative of the Cape farmer to utilise his Fynbos land profitably cannot be overlooked. To the Fynbos owner, the financial principle understandably often outweighs other considerations.

The major part of the original Fynbos-area usable for intensive farming has been brought under cultivation of a wide range of historically established crops (+ 60 per cent). Presently only small, marginal areas (less than 5 per cent) remain as potentially cultivatable land. These might be utilised for the production of special crops, for example wildflower cultivation. The rest, perhaps 30-40 per cent, is uneconomical, mostly mountainous terrain of which parts could ideally be conserved and managed as a relic of the original Fynbos system.

In increasing order of financial profitability per unit area the present-day, privately owned Fynbos areas may be

- left undisturbed — these constitute a very large area
- regularly burnt to obtain young palatable growth for grazing purposes — a well-known, extremely destructive practice to Fynbos involving large areas
- picked for fresh and dried flower marketing — constituting large areas
- cultivated for wildflower production — these constitute small areas, probably less than 5 per cent of the present-day remnant Fynbos as explained above.

The most efficient and limited use of Fynbos today is therefore probably for the purpose of intensive horticultural production of wildflowers. Orchard cultivation of especially proteaceae may relieve areas of veld estimated at ten to one hundred times larger from the pressures of picking. This may be attributed to increased yields per plant as well as higher densities of plants per unit area. In addition the value of cultivated material is higher, due to its improved genetic and cultures qualities (for example less damage by pests).

(Of course a significant portion of the traditional agricultural crops may in addition be replaced by cultivated wildflowers, as has been happening on a limited scale.)

From the financial viewpoint as well, a study of the requirements for the optimal development of the intensive wildflower production industry is therefore justified.

Development of a strong wildflower export industry based on intensive horticultural production.

Relative to its small size the present wildflower industry has claimed a respectable 1 per cent, approximately, of the South African horticultural export economy. As such it contributes significantly to the earning of foreign exchange and therefore to the national welfare. In addition the industry is labour intensive and is thus a good job provider in the Cape Province. It is especially the *intensive* wildflower production industry that bears promise of increased and improved provision of unskilled and semi-skilled jobs in the future.

The wildflower export industry has risen to a unique leading position in the production of especially protea cut flowers for the European flower markets. Significantly there has been a rising international interest in protea cultivation during the past decade as is shown by the fact that at least seven countries now produce proteas (Australia, Bolivia, Israel, New Zealand, Portugal, U.S.A., and Zimbabwe).

The export industry can improve its economic position and its international leadership in the production of protea cut-flowers and other wildflowers. These expectations are based on an average annual growth rate of approximately 20 per cent for the industry in the 1970s and the fact that less than 1 per cent of the European fresh-flower market has so far been penetrated!

An enlarged and more stable pattern of demand and supply must however be developed in the industry's interest. It is *especially the intensive production industry that will have to be developed* to attain these objectives. Unfortunately the basic knowledge needed to direct the development of intensive production is lacking.

The main questions on which information are needed are as follows:

- To what extent are wildflowers produced in orchards of picked from the veld?
- What are the regional areas suitable for the optimal production of specific varieties?

- What is the potential for industrial expansion at the expense of (privately owned) Fynbos and existing agricultural land?
- What species presently picked from the veld may be produced more economically in flower orchards?
- What are the growth rate of and present trends in orchard cultivation?
- How effective are the production methods utilised in the industry?
- To what extent can the general poor quality of cut-flowers and the leaf blackening problem be ascribed to ineffective harvesting and post-harvest handling methods currently used?
- Should the present export quality requirements, based in veld production standards, be revised?
- How serious is the need for genetically improved material?
- How labour-intensive is the industry and what are the major problems?
- Do the present practices of individualistic marketing and opportunistic veld production jeopardise the long-term interests of the industry?
- How may the intensive production industry improve its organisation and its recognition as a fully-fledged branch of agriculture?
- What are the attitudes and expectations of the intensive wildflower producer?

Some of the industrial relationships which need a better understanding are as follows:

1. What is the economic relationship of orchard production to veld-picking?
2. What is the relationship of fresh — to dried-flower production?
3. What is the potential of wildflower production relative to the rest of agricultural production in the Cape?
4. What is the relationship of present seasonal cut-flower supplies to market demands?
5. How may the relationship between (orchard) producers and exporters be improved?
6. What official guidance do producers need?

These questions are all facets of the same general problem, viz. a lack of basic information on intensive wildflower production. It is therefore desirable that a study of the scope, potential, and problems of the intensive Cape wildflower industry be launched with a view to promote its continued growth and to improve the basis for a Fynbos conservation policy. An industrial survey by means of the questionnaire method is probably justified.

The industry may approach one of the several independent (agro-) economic institutions which may be interested in undertaking this type of study.

Walter J. Middelmann

APPENDIX G:

Marie Vogts Appointed As Researcher at Fruit and Food Technology Institute (FFTRI) Stellenbosch 1964

Planned Research on Indigenous Flowers

Actual planned and continued research dates back to 1965 when first Dr. W. Horn and later Miss J. Bundies were recruited from Germany to give special attention to breeding research. Although a start was also made with proteas, the main emphasis was placed on certain indigenous bulbs. Considerable progress has been made. Other technical sections of the Institute have, as before, done part-time work on insect pests and cultural problems, mostly in the deciduous fruit off-season, and in this way gradually laid a foundation for the great new commercial horticultural industry of the Western and Southern Cape, which is attracting considerable interest and developing rapidly.

The transfer of Mrs. Marie M. Vogts, the well-known specialist on the growing of proteas, to this Institute towards the end of 1964, provided a new stimulus to commercial growing of indigenous flowers. Apart from the specialised advisory service which has become available to prospective protea farmers and others, Mrs. Vogts was made available to the Department of Forestry to assist with selection of seed collection plantations in Government Forest Reserves. By this means the Department of Forestry could make available to growers and the public selected seed of Proteaceae on a large scale at reasonable prices.

During the past few months, this Institute has concentrated attention on certain fundamental obstacles, amongst other vegetative propagation of Proteaceae and critical problems in connection with keeping quality and export by ship to the Northern hemisphere. Very promising results have been obtained already. However, there are still many problems to be solved and more intensive experimentation is necessary before large-scale shipments by sea, which is much cheaper than by air, can become a practical reality. Almost all overseas countries are intensifying their phytosanitary requirements in respect of import of plants, plant products, and even flowers. This implies that strict inspection standards will have to be complied with before the necessary phytosanitary and export permits can be issued. Farmers who wish to plant indigenous flowers for export will, as is the case of fruit and other farmers, have to take the necessary precautionary measures for production of high quality products free from disease and pests.

Facilities for Research on Indigenous Flowers

At its headquarters in Stellenbosch and its main experiment station 'Bien Donné' in Groot Drakenstein, the Institute already has a series of well-equipped laboratories, experimental cold stores, special facilities for the study of insect pests and diseases, soil studies and trace-elements, glasshouses with air-conditioning, etc.; in short, for all research which is necessary in connection with propagation, culture, keeping quality and transport, genetic studies and breeding of indigenous flowers.

For experimental plantings and cultural studies space has been set aside on seven existing experimental farms and in experimental plots for this Institute, which stretch from Citrusdal to the Langkloof, that is, over a range of soil and climatic conditions. Agra-meteorological data are collected continuously on the experimental farms and plots. In cooperation with the Forestry Department, provision has been made for experimental plantations at Bettys Bay and Kleinmond (Oudebosch) to meet certain special requirements. There are also facilities available for experimental plantations on certain of the experimental farms of the Stellenbosch-Elsenburg College of Agriculture of the University of Stellenbosch.

The provision of additional personnel for full-time research on the problems of the commercial indigenous flower industry is being considered seriously.

Help from and Cooperation with the FFTRI

The Institute already has a considerable amount of information in connection with commercial growing of proteaceae and some other indigenous flowers and can supply advice upon request by letter or during personal visits at the Institute at Stellenbosch on the visitors' day, namely Thursday.

It is known, and has been shown clearly in surveys, that the species of Proteaceae and Ericaceae hybridize easily in nature. The natural hybrids are largely inferior. Now and then one might occur which has one or more characteristics, which could be of economic importance either directly for commercial growing or after further planned breeding. It is therefore of national importance that hybrids which have exceptional qualities should be preserved; here the Institute can be of help with its special facilities and particular techniques on a quid pro quo basis.

APPENDIX G:

Research into the South African Proteaceae:

In the Beginning . . .

The year is 1940. Encouraged by beautiful protea blooms on plants I cultivated in the mountains and sorely discouraged by failures elsewhere, I am filled with an urge to find out why. No book on cultivation or propagation is of any help and I nervously approach the highest authority. '*Young lady,*' Professor Compton says, '*you are wasting your time. The backlog on this kind of knowledge of the Proteaceae is too great. Forget about exploring reasons for their behaviour and their hidden characteristics and don't try to bring high mountain proteas down to open flats*'.

This was the challenge. Forty years ago there was no scientific guide-line to explain the causes of the eccentric behaviour, reactions and needs of these plants. The only attempt to present classified knowledge was made by Joseph Knight in England almost two centuries ago. But as his work was limited to cultivation under highly artificial conditions, it could not serve as a basis for continuation of research and interest soon waned. In South Africa, only a few hints and odd bits of information could be gathered from botanic gardens and from the few individuals who had successfully cultivated single plants. Even these limited recommendations and inferences were based on reactions of proteas in the particular places where they had been cultivated, and these were all in the winter rainfall region. No one had been interested enough to spend time and money on investigating the causes of the obvious divergence from known horticultural crops. Nevertheless, I took up the challenge and launched a new approach to the problem, different from the age-old standard trial and error method of accumulating preliminary knowledge through practice.

In the beginning of this scientific research, the natural habitat served as a reference book. The apparent similarity of reactions to a number of factors of many species was so encouraging that the family was treated as a whole (in spite of a few glaring exceptions) and eventually 52 species, representing eight genera, were used for intensive investigation. I made numerous observations, which seemed relative to the project, first in nature and then, by experiment, observed under control. Since this research had to be considered as an undeveloped branch of science, analogy played a prominent role in the classification or grouping of what had been perceived. Analysis and synthesis, as far as possible, had to be applied to discover some kind of order in the vast number of and seemingly confused phenomena. It was necessary and fruitful to make tentative suppositions even though the hypotheses sometimes turned out false.

Any progress in the preliminary investigation would have been difficult, if not impossible, had observations not been controlled in both winter and summer rainfall areas. Similar reactions could be sorted out and were accepted as part and parcel of the characteristics of the various species, while others were clearly due to environmental conditions and influences.

Towards the late 1950s a good deal of knowledge concerning cultivation had been acquired. It became clear that by following certain guidelines the growing of proteas in different areas was a possibility and that cultivation was not restricted to a few mountainous places in the Cape. Public interest was so stimulated that I was compelled to publish my findings — the first book in South Africa on cultivation: 'Proteas Know Them and Grow Them' (1958). This was the initial breakthrough in cultivation possibilities. Attention was immediately focussed on the economic potential in a world always clamouring for something new and different and thus also for new and exceptionally beautiful cut flowers.

However, the aim of probing the underlying causes of reactions and the hidden characteristics was far from achieved. For this reason, and also spurred on by the general wave of general interest in the commercial potential, in 1960 I made a public plea for more research — with remarkable results. State money was granted and research on proteaceae was undertaken by an increasing number of scientists. The research grew from strength to strength. It is clear that had it not been for possible commercial value, research, essential also for conservation of the original sources in the veld, would still have been sadly lagging.

With the increasing interest in commercial growing an overall survey of the Cape Proteaceae in their natural state had to be undertaken in order to select, from the 300 odd species, those with economic potential. I discovered that the species with the highest potential consisted of several variants (apart from ecotypes). The apparent differences and inconsistencies within a species had long been a tantalizing phenomenon. Not only flowering time, but other distinctive qualities such as stamina, seed production, resistance to pests, reaction to nutrients and, above all, ease of cultivation, seemed changeable and unpredictable. But since nothing in nature is haphazard a thorough and systematic investigation was imperative. The results were illuminating and have had a far reaching effect on the commercial production of proteas.

Supported by two State Departments I undertook this formidable variants directed survey from 1962 to 1972. The area covered was the natural habitat of the Fynbos Proteaceae in the Folded Mountain Region of the Cape, from the Cedarberg Range to the Amatolas. Localities on the higher slopes of the mountains received prior attention, since it was here that populations that had been left relatively undisturbed through the ages could still be found. Due to the highly dissected topography many populations had been completely isolated. Through evolution, specific although varying characters of the same species had become dominant in the different localities. This resulted in spectacular dissimilarity.

Microclimatic conditions, as well as soil analysis of some localities, were seemingly so alike that the term *ecotype* was discarded. In its place the vague term *variant*, and subsequently commercial variant, was adopted. Economic value, in regard to attractive appearance demanded by the trade, as well as horticultural potential, was evaluated. Perhaps the greatest contribution to commercial growing was the discovery that many outstanding and worthwhile features, including flowering time, remain stable (tested up to the fourth generation), when cultivated far from their natural habitat, even in the northern hemisphere. Variation is a remarkable asset and is being exploited to great advantage, e.g., variants of *Protea cynaroides* or those of *Protea repens* flower in different seasons. It is thus possible for the grower who can successfully raise these species to select variants offering *P. cynaroides* or *P. repens* flowers, either right throughout the year or in the seasons most profitable for his market.

This finding of the preliminary investigations of the many peculiarities provide basic material for further research and only now, after forty years, many one dare to assess the value of the initial steps which made advanced and vital research on the domestication of the South African Proteaceae possible.

Marie Vogts

25 Sept 1984

APPENDIX I:

DEVELOPMENT AND STRUCTURE OF THE SOUTH AFRICAN PROTEA INDUSTRY

A talk prepared for the International Protea Conference to be held at
Kallista, Vic., Australia, 4-8 October 1981.
Walter J. Middelmann, South Africa

Let me start by saying that for us, coming from South Africa, it is gratifying that Peter Mathews' Conference is called an International Protea Conference.

We all work with proteaceae. You in Australia have a good many more of them, species-wise, than we have, but it is the protea proper which has more and more made the headlines. So much so that your Waratah has become known as 'the Australian Protea' and that Banksias, Dryandras etc., are often grouped in the flower trade with the Protea proper. Your own Conference Program lists them that way.

You have gathered here essentially as commercial growers aiming at the cut flower trade, probably largely in the end, export orientated. You want to listen to the experts how to go about it. Much has been heard during these two days about the technical side. Now I want to give you something of how things developed in our country: how from small beginnings an industry with something like an annual turnover of Australian Dollars 6 million has emerged. There may be parallels and there may be lessons in this for you people, though there are also some very essential differences.

It is hard for somebody not really acquainted with the Australian scene to bring comparisons or to guess what aspects you would be most interested in. This is evident: there is great enthusiasm amongst Protea Growers in this country; otherwise you would not be here. Surely there are many questions you want to have answered and probably more will arise after I have told you my story. During question time and in discussion I shall do my best to answer these, but there is one recipe nobody can give and that is how to make instant money out of proteas. Yet, I have mostly found that, human nature being what it is, that is the crucial motivation.

Anyway, let us start with the history, with how it all came about.

The wealth of the South African in particular that portion known as the 'Cape Floral Kingdom', comprising of what botanists call a Mediterranean Macchia type flora, is well known. We in South Africa call it the 'fynbos' (fine bush). Some of its items have in fact been commercialized for a couple of centuries or longer, mainly by the Dutch. Their seafaring captains and traveling discoverers brought specimens home. From these breeding took place, firstly with bulb species such as Iris, Gladiolus, Freesia, but then also came Pelargoniums, Geranium, Agapanthus, Gerbera, Strelitzia, some Ericas, and many others. All this work was done outside South Africa, as within the country only a small population of European descent and potentially flower-conscious people existed. Their occupation mainly as isolated farmers, 'boere', did not allow much time for cultural pursuits and thus offered no demand and no market.

Only in the present time has commercialization inside the country set in, say during the last ten to twenty years, namely that of the South African Proteaceae. In the book called 'Exoticorum' by the Flemish botanist, Clusius, which appeared in 1605, a dried head of *Protea neriifolia* was the first ever South African floral item to have been depicted and today hundreds of thousands of these are sent into the world, fresh and dried. Collectors later brought seed to Europe, maybe even small plants, and in the beginning of the nineteenth century, collections of South African Proteas flowered in various greenhouses in England and on the Continent. Joseph Knight in 1809 produced a book on their cultivation. It became a kind of status symbol to posses them among the garden-conscious gentry in England.

Then came the hot, humid orchid and palm houses and the proteas died, since they needed, at the most, a cool glasshouse for protection against frost but otherwise love the light, wind, and fresh air.

South African Ericas though, were kept and propagated, specially in Vienna, and their direct descendants can still be seen in Schoenbrunn and Belvedere. I will not say more about Ericas except that hardy hybrids were developed, with South African as well as European parents, mainly I believe by Italian gardeners. Today there is also a vast Erica industry flourishing in the San Francisco area, California, for fresh-cut flowers as well as for pot plants.

In South Africa, certainly nobody took an interest in the protea. They were just a part of the natural shrub on the mountain and only the coloured hawkers stripped some of the bushes, haphazardly, and sold blooms in the streets of Cape Town for those who could not afford 'proper' flowers. The ordinary householder cultivated roses, chrysanthemums, carnations, and many other European annuals in his garden and his house would normally be decorated with these.

By 1913, the National Botanic Garden at Kirstenbosch were started just outside Cape Town, with the idea of exclusively studying and cultivating the native flora of the country, to make the public conscious of its value and to spread knowledge of cultivation practices. It took some three decades before the first gardeners followed. The Botanic Society of South Africa pioneered the Flora Conservation movement in regard to protection. In 1937, a provincial ordinance was laid down with certain rules to achieve conservation of the native flora. Gradually a few enthusiasts collected seed and created gardens of natural plants, mainly at the Cape itself.

The biggest boost came from the efforts of the late Frank Batchelor who, from 1944 onwards, built up a wide collection of showy bushes on a hillside on his farm near Stellenbosch. He invited the public to visit there and arranged for exhibitions of his flowers at leading Department Stores. This created a storm of interest and a sudden demand for protea flowers from the public and then from the florist trade.

From just before this time our personal interest and history started. My wife, Ruth, came to Cape Town from Johannesburg in 1940 when we got married. She saw the proteas on the mountainside around where we lived, high above the Atlantic coast in one of the seaside suburbs of Cape Town. Endowed with a 'green thumb', she collected seed and raised seedlings. We ventured further and further afield, collecting a large number of different species and variants from all the mountain ranges of the Western Cape and even further away. With many setbacks she developed her technique and a collection was built up. People came and saw this and asked for plants. Thus a specialist 'backyard nursery' was born.

In 1947, we bought a large piece of land called 'Honingklip', 60 miles outside the city, essentially as a weekend retreat. It had a good bit of natural proteaceous flora, but also lent itself to planting and introducing of other species. No staff was initially employed nor machinery available, yet years of weekend work created the possibility of supplying florists in Cape Town and further away too, with both cultivated and 'wild' proteas and pincushions (*Leucospermums*). The plant nursery endeavored to offer the widest possible range of the species of the family. We were of course, soon not the only ones. Others followed and did the same, some on a very much larger scale eventually.

Another pioneer, Dr. Marie Vogts, in the wake of all this, published the book, 'Proteas, Know them and Grow them' in 1958, with advice on cultivation. Yet, no research had so far been started other than that at Kirstenbosch, which was not really directed at commercialization, but rather at popularization. In the early sixties, an exceptionally cold winter occurred in Europe. Air transport had already started supplies of carnations and roses from the glasshouses around Johannesburg to European markets. Buyers from Europe, looking for supplies of flowers then so-to-speak 'discovered' the protea. Suddenly there was talk of 'millions'.

High hopes developed, but European buyers soon found that the proteas' main flowering period is in our winter, that is, their summer. Flowers in Europe are cheap then and plentiful while a large part of the population is on holiday. The Europeans wanted Proteas from September/October onwards, with a peak at Christmas, and could not then understand that these exotic things were not readily available at all times. Even today of course importers cannot get any single item, except perhaps *Protea cynaroides*, the year round, something which for 'normal' flowers is a matter of course.

It took a long time to overcome this problem, since anyway practically all the exports originally came from the wild, under license from the Cape Nature Conservation Department.

However, exporters looked around and seizing the opportunities, found that in the Western Cape Province there were actually, a little further away, variants which do flower in the Southern summer. This is mainly due to climatic factors, namely in the higher, cooler regions, though some are even genetically fixed such as the summer-flowering variants of *Protea cynaroides* coming mostly from a good deal further East, and also the red Protea repens from the Eastern Cape. Exporting thus became a collecting operation. No landowner, nor for that matter grower of varied introduced cultivated species could hope to just work from his own material alone. It was hard for the Dutch and German importers to understand this when greenhouse growers elsewhere could produce almost anything in one and the same place and at any time. Here the exporter had to explore sources in far-away and even difficult to reach areas. He had to arrange for transport over hundreds of miles — he had to convince farmers that flowers had to be picked regularly, for fixed collection dates to coincide with air transport, come rain, wind or snow — something which many of these farmers just were not used to. It must be stressed here, by the way, that state bodies such as the Forestry Department in particular do not allow picking on their open spaces. All material picked in the wild comes from privately owned ground, under license.

It took the exporters years to develop supplies at our end and, in collaboration with mainly Dutch and German importers, also years to initiate the marketing overseas. However, they got the public to accept these 'exotics' and to create a continent-wide regular supply right into the flower shops of the smallest towns in Central Europe.

There are many problems. The original supplies from the wild, from the 'veld' as we say, were, and are, often rather scruffy, varied, stems too short, crooked stems. Foliage is largely insect-damaged and blackens easily, especially under higher temperatures, which is a peculiarity of this family. Actually most of the leaves are stripped off selectively for shipping.

Plantation-grown flowers are more uniform, in the case of *Leucospermums* usually have larger flower heads, and generally are of better appearance. Yet up to recently, having all come from seedlings, they too show tremendous variation. Research was recognized as a necessity already by 1963 even though exports then were still minimal. A Protea Research

Unit was created under the aegis of the Fruit Research Institute in Stellenbosch, where under Dr. Vogts' direction, pioneer work was done. Actually in some ways, such as seed germination tests, vegetative propagation, selection, hybridizing, people in New Zealand like the late Wallace Stevens, Duncan & Davies, or the University of Hawaii, initiated research areas, which antedated what was done in South Africa, but then conditions differed too.

Today this research is in full swing, under the direction of Gert Brits of the Protea Research Unit now under the Director of Horticulture, also by Dr. Gerard Jacobs at the Department of Horticulture of Stellenbosch University. The latter's work is done at Protea Heights, the original Batchelor farm. Growers are now applying results of research and a number of them now work with mist spray units and select good forms from their own material to which they give fancy names. They also work from early or late-flowering individual plants in order to extend the flowering seasons. For two years now, the first selections of uprooted cuttings have been released by the Department of Horticulture.

It should be mentioned that in a country where there is still the possibility for some people to do large-scale collecting of seed heads on various privately owned mountain areas it is possible, in addition to orchard-like systems of row plantings, to have unsorted seed broadcast-sown over big prepared lands and part of present flower supplies come from such fields. In any case, most proteas in South Africa are not grown under irrigation, contrary to those in California or Hawaii. The eventual development will undoubtedly tend towards the orchard type plantations, with selected, vegetatively propagated, uniform material of best quality where, hopefully, every single flower can be usefully and easily harvested. Cooling facilities should be provided in the packing shed on the farm and packing should take place right there to avoid damage in transit.

Consignments should then go direct to the exporter's collection depot at the airport. It is a big contrast to the original method of sending labourers up the mountain with bags, paying them by the piece, taking the material down to the shed, having to throw half of what was collected away because of physical damage of because of poor appearance, and then perhaps carting roughly packed flowers for hours in the day's heat to the exporter's premises, there to be re-graded, bad foliage removed and finally packed for shipping.

Undoubtedly the properly equipped 'scientific grower' will win the day.

I have tried to show you how the South African Protea Industry developed, since to most people in the world's flower trade this experience is totally foreign, almost inconceivable. In Australia there is of course a partly comparable situation, namely where Banksias and Dryandras in particular come from the wild for export packing, fresh or dried. One can also compare the collection of greens in die forest areas of Washington State as another example of utilizing a natural, uncultivated natural resource.

Now to give you some figures:

In the sixties, South Africa's total flower exports stayed around the R800 mark; say about A$.750 The Australian Dollar is near enough to the S. A. Rand, so just equate my Rand figures with Australian $ for ease of comprehension.

Maybe this figures included a couple of thousand Rand's worth of Proteas — the rest were carnations, chrysanthemums, roses etc., mainly from glasshouse culture around the airport at Johannesburg. From 1970, this took off and what made it take off was the proteas. By the time the total exports reached R2.5 million in 1975, half of this was proteas and by 1979/80 the value of proteas exported reached R2.5 million by itself, fresh flowers only, though including the so-called 'Cape Green'.

In quantity this means about 1.700 metric tons or 1.7 million kilograms, quite good business for the airlines. You will want to know about price. The prices I quote now are F.O.B. Cape Town Airport. For the medium size proteas, say Compactas, Neriifolia, and Obtusifolia, the price is about 13c to 15c per bloom. *Protea magnifica* (the old name was Barbigera) sells at 40c for the small one and 60c for the large ones, while Cynaroides fetches between R1.10 to R1.30 and Grandiceps sells as 40c. Pincushions i.e. the *Leucospermums* are more subject to rapid variations in price. Gluts occur since the flowering periods are shorter and large masses of blooms are suddenly offered by producers all at one time. Thus early and late Cordifoliums could be up to 15c, dropping in the main seasons to only 10c, while price cutters sometimes go even below this. *Leucospermums patersonia* is less popular, but starts earlier and would go at 7 to 9 cents.

Greens, which are usually bunches of various foliages, Heaths, Brunia etc. are sold by the carton, 25 bunches to the carton, mostly at R10 to R12 per carton, with some special items realising slightly higher prices.

It is an unfortunate fact that these prices have remained unchanged for the last few years, inflation notwithstanding. Your producer at the farm will as a rule obtain about half the prices just mentioned for the flowers supplied by him loose. These would then have to be packed by an exporter. This looks like a margin of 100 per cent but it is of course not so, since transport costs, cartons, wages, losses during final packing etc. must be considered. The exporter is lucky if he can put on 30 per cent gross profit out of which he must now pay his office, telex and shipping costs, salaries and so forth.

In the course of time three distinct types of exporters have emerged:

- The producer cum exporter, growing part of his material himself, based in the country, doing his own packing and that of bought-in material. All exporters must operate by buying in over wide area hundreds of miles away sometimes, to obtain a wide selection.
- Then there is the
    - exporter/packer, mostly situated near the airport, who partly collects, partly gets delivered his material from all over, and finally the
    - pure exporter, just operating an office, again at the airport, who only buys ready — packed material, in the export cartons, prepared by the most reliable growers, partly delivered to him and partly collected from growers.

It is obvious that, as a result, variations in prices between the different parts of the chain must exist. Everybody in the export game must be always on their toes. The transactions are done by overseas telex or telephone. Demand changes from day to day, often influenced by weather changes in Europe. There is a lot of night work, certain planes must be met, and schedules are changed. Availability of air spaces varies, and it specially becomes difficult near Christmas. There are risks that flowers must be transshipped and stay over in some airport such as Rome or Madrid instead of going to Amsterdam or Frankfurt.

While the main German importers will see to distribution within their own network of wholesalers in Germany, the Dutch ones sell throughout Europe. They all sit at or near the flower market at Aalsmeer, but hopefully will not put any of their imports on to the auction. Only at time of glut a part of the imports may get auctioned and, as a rule, this leads to a complete price break. Normally all proteas are sold by the exporter against a firm order and this calamity, which then hits the whole trade is avoided. It must be clear from all I have said that there is considerable competition within the trade, as there is between the European importers. This keeps prices down to the consumer, but also means ever lower margins in the trade.

Coming to the airfreight factor, we all know that during the past eighteen months in particular, and worldwide, airfreight rates went up several times — in addition, the Rand appreciated against the Deutch-Mark and Dutch Guilder, which in itself increased the cost of imports to the Europeans. The results were devastating as far as far as the protea export trade is concerned. Total exports were down 11 per cent in 1980 as against 1979, and a further 10 per cent during the first half this year. Finally, we ended up with R1.90 per kg for airfreight by July. Now the airlines and South African Airways in particular woke up and, following high-level representations by the industry, we have had a major reduction as well as a 'co-rated tariff' between Cape Town and Johannesburg, which in fact means that it costs the same from either airport to all main European destinations. Only when the airlines found that they went North with a lot of unsold freight space did this happen and one just wonders whether at times of high demand and full loads, say mid-December, they will stick to it.

Just to make all this more realistic, the relationship between the value of flowers to airfreight involved had been 50:50. It went so bad that it came to 40:60 against the flowers and we were selling airfreight rather than flowers!

One consequence was that experiments were made shipping proteas by sea in refrigerated containers, at 4° Celsius. Technically it is possible, but the slightest temperature change causes condensation and almost total loss. Only part of the 'greens' arrives really well. Also, container shipping has not got the elasticity of air transport. Too much arrives at an uncertain date at one place. Thus it seems unlikely that it will become of greater importance.

One problem remains a major one for the exports, namely insects. We are blessed with a multitude of creepy-crawlies, which have proved extremely difficult to keep in check, the more so since we are situated in the natural home of these things which makes it quite impossible to control them in the field. Fungus troubles must be added. Even post-harvest treatment so far is quite imperfect, despite ongoing research. Methyl bromide, the only real treatment, spoils the blooms. Your Department of Agriculture would never admit our flowers, nor would the largest potential market outside Europe, namely North America. This, by the way, will be a great advantage for you as potential exporters since your material will be a great deal cleaner than ours, and also will have better leaves.

Now a few words on the so-called Dried Flower trade. Peter Matthews originally had the title of my address down as 'Drying Proteas', probably because our own main enterprises are the production and export of natural dried floral materials. Here too the first shipments started in the early sixties, first to Europe, but today to quite an extent also to the United States, Canada, Japan, and a little even to Australia. We have a wealth of fine ornamental items outside the protea family too, such as the grass-like Restionaceae, the Brunias, Thorn branches, stalks of Watsonia, Aristea, and not to forget the Helichrysums (also known as Everlastings) and known in Australia as 'silver daisy'. We even have good use for the woody, beautifully shaped Hakea pods, which, originating from Australia, are in fact a pest plant in South Africa, and are proclaimed weeds. We are not to forget a variety of fine Eucalyptus pods either. However, again the proteaceae offer us the widest range of utilization, from slightly faulty rejects of the fresh-cut flower trade hung up for drying, to the vast range of woody seed bases and stars, specially *Protea repens* and *P neriifolia* in all their forms. On our own list we have no less than thirteen products derived from Repens and twelve from Neriifolia, and then some of these bleached as well. So you see it is not only a matter of drying proteas! As far as the techniques used, the latter are mostly simply hung up and air-dried, often with treatment in a sulphur chamber immediately upon picking, which improves their colour; keeps them brighter. We also employ methods of chemical bleaching and preservation of foliage with glycerine solution. A certain amount of dyeing is done by dipping as well as by absorption.

Total exports of this part of the industry are estimated at R2 million. I had mentioned a similar figure for the fresh-cut flowers and we have estimates of 1.5 million Rand for the local market. Total then would be 5.5 to 6 million turnover for the whole industry of protea and allied floral items, of which at least 75 per cent is exported.

I should mention here that with some difficulty I managed to obtain Australian figures and find that, other than orchids, your total exports of fresh and dried flowers and foliage came to A$. 1.570. Of this $502. went to Germany, $435 to Holland, $169 to Japan and only $70 to the U.S.A. There is no means of establishing how much of this might be from the Australian native flora. I am inclined to think that the latter two countries, i.e., Japan and the U.S.A. will be the main future outlets for your protea exports. It seems unlikely to me that you would be able to compete with South Africa in Europe, except on quality and freedom from blemishes caused by insects and fungi. Your seasons are the same, your distance too far. But North America and Japan are a different matter.

Now to Propagation. I gather a good many of you here are nurserymen intending to supply flower growers with plants. Others will be flower growers who propagate their own plants. Here again I want to outline how things developed in South Africa. Well, when the whole movement started, we ourselves, and others, thought that a lot of private plantings in gardens and flower plantations would bring with it rewards. Actually, after initial enthusiasm, sales later went down and the proteas still play a rather minor part in the big nurseries' sales. There are of course large parts of the country climatically and/or soil-wise unsuitable, though in the Transvaal, actually proteas can do very well until sometimes frost or hail hits them, which is not the case in the Cape. A few very fine commercial plantations have developed not far from Johannesburg. Today a few specialist nurserymen serve the demand of the private gardening sector while Kirstenbosch and the Botanic Society continue, also by means of annual plant sales, to encourage the general public.

Flower farmers mostly do their own propagation since transport of container grown plants, in addition to their cost, is often prohibitive. However, once specialist nurseries might be able to offer selected, vegetatively propagated material of outstanding characteristics, this position might change and it should become worthwhile for the farmer to acquire such material. The Protea Research Unit has, in 1980 and 1981 started releasing selected cultivars for further propagation. A small export of unrooted cuttings has also started.

Seed is available from Kirstenbosch and from a very few private suppliers. Without blushing I might state that our own enterprise has played a major role in protea seed exports to all suitable areas of the world, including Australia, for over two decades. Careful sorting, all by hand, feeling every single seed, is required. A wide range of sources and constant replenishment with fresh stocks must be arranged for and experience is not easily gained. Seed at present offered by major dealers worldwide comes mostly from us. With the great demand that has arisen here in Australia it should be borne in mind by buyers that the resource is limited, particularly with some items like *Serruria florida* and *Protea speciosa*. Plantation-grown proteas will have been picked for their flowers and pruned at season's end. Thus, most seed must be collected in the wild and the further you must go to find some, the higher will be the production cost. Thus ever larger orders are ever harder to fulfill and quantity does not make the commodity cheaper.

Let me still touch on the matter of Nature Conservation and the Wildflower Trade with respect to the utilization of the native flora. When things started, well-meaning though not very knowledgeable people spoke of 'the rape of the veld'. The idealist, mainly townsmen to whom 'nature' is part of their weekend recreation, could not stand the thought of using the natural flora for the production of 'filthy lucre'. This was desecration! The farmer, on the other hand, lives with nature all the time. His job is to make a living off his land so he tended to regard the natural vegetation often as 'weeds', unpalatable to his animals and just in the way for cultivating crops. Where he could, he would actually burn it off plough it out.

Now, however, in many areas this has changed. The 'weeds' have become valuable. They can be exploited to bring a return and the farmer now looks at them as an asset. There has undoubtedly been a certain amount of plundering, one might say 'mining' of the veld in the first years. There is still a lack of knowledge of good management, correct pruning, leaving a reserve for re-seeding, when to burn for regeneration etc. By and large, however, a store of knowledge is being built up. Let me finally quote from a recent statement by the Cape Department of Nature and Environmental Conservation, which sums up the position very well:

> *As far as the principle of the wildflower trade, or game farming for that matter, is concerned, it must be pointed out that the wise use of renewable natural resources is an accepted form of conservation management. The problem starts when overuse, either through ignorance or greed, affects the stability of natural population. Obviously the Department would support any move to restrict or prohibit the use of any rare or endangered species.*

A further word about the trade itself. We founded, in 1966, an organization called the S.A. Wildflower Growers Association (SAWGRA), to represent the protea industry towards Government, Railways, Nature Conservation Department, Airways, and the public in general. At that time plant nurserymen played a major role. Parallel with the developments already described, it was the flower producers and exporters who soon became dominant and a change of name to S.A. Protea Producers and Exporters Association (SAPPEX) indicated this. We have over 200 paying members now, an active Committee and a fine quality quarterly Newsletter. Membership is open to all interested persons, also in other countries. SAPPEX subs are R25 per annum, plus R5 for foreign postage, while subs for Newsletters only would be R15. SAPPEX members cover probably 90 per cent of the export turnover of the industry and the voice of SAPPEX is recognized by the authorities, although maybe not always listened to. SAPPEX has for instance negotiated on matters like that of grading and inspection of export flowers and has played a major role in for instance standardization of cartons and naming of new varieties. It has produced a brochure, which would also be of use to you in Australia. The organization is in regular touch with Union Fleurs, an organization which represents essentially the European Importers and which has its seat in West Germany. Most of the exporters personally visit the overseas markets as this contact is of utmost importance.

At present there are five main fresh-flower exporters and at least another five more minor ones. This leads to strong competition on the buying side for the rarer items in the season, and of course, also on the selling side overseas. Naturally there are conflicts between producers' and exporters' varying points of view in some respects and certain producers feel that some sort of Cooperative would 'eliminate the middlemen'. In fact, however, the wildflower export trade where long and odd hours are the rule, where quick decisions and elastic methods have to be applied, is a very personalized type of exercise. So far the individual, almost family-type enterprises have done this more efficiently than the overhead-plagued larger concerns. There have been a few casualties in recent years and it is not surprising that the most recent one was the one with the highest capital injected into its establishment.

Well, I have tried to give you a wide canvas of the development of the Protea Industry in South Africa. We think ours is quite a success story, but, like everybody, we also have lots of problems. One thing is sure, however, the protea globe, of which this Conference is proof. Proteas are 'in' at many places as one might say. Yet, as it was stated during a lecture at Brooklyn Botanic Gardens, New York some years back, 99 per cent of people in the United States would not know proteas from proteins! And Jacob Ben Yaacov in Israel made a remark that 'Proteas are where Roses were a thousand years ago', though of course with present knowledge and techniques, the development would not take anywhere as long.

APPENDIX J:

The Conservation of Genetic Resources in the Southern African Proteaceae
JP Rourke at Zimbabwe IPA Conference 1993

'Cherish variation, for without it life will perish' — *Sir Otto Frankel*

What I have to say to you today concerns the continued growth, development, and future survival of the protea cut-flower industry. As growers you are faced with practical problems concerning very pressing issues such as marketing, promotions, sales or air freight tariffs, and, naturally, these will be uppermost in your minds. Understandably therefore, the conservation of genetic resources will seem a rather esoteric subject of little immediacy, but it does have medium to long term economic implications and consequently, it behoves us to give careful consideration to this matter.

The domestication of proteas is a very recent phenomenon. It began in the late 1940s at Stellenbosch when Frank Batchelor began selecting deeper coloured forms of *Leucospermum cordifolium*, followed a few years later when Jean Stevens in New Zealand started deliberately hybridising *Leucadendrons*. Batchelor's twenty-five years of work culminated in the production of a deep red *Leucospermum cordifolium* selection registered as 'Mars' in 1972. On the other side of the Indian Ocean, a hybridisation programme with Leucadendrons was initiated in the early 1960s by Jean Stevens who crossed *Leucadendron laureolum* with a red-leafed form of *Leucadendron salignum,* which her husband Wallace had collected in the Langkloof area of the southern Cape. Among the seedlings resulting from this cross was the renowned *Leucadendron* 'Safari Sunset'.

Since then, hybridisation and selection programmes have continued apace resulting in the registration of well over 250 cultivars in various genera indigenous to the Southern African region. Thus within the space of our lifetimes we have been witnesses to the commencement of processes that are resulting in the domestication and improvement of a group of wild plants to form a whole new horticultural crop. By comparison with roses, potatoes, carnations, or wheat, which have been cross bred and selected for thousands of years, the proteaceae are only just on the threshold of being manipulated for the specific needs of man. This is a promising start, but if the industry is to survive, breeding and selection will have to be intensified in the future.

However, if we are to continue with domestication and breeding programmes there are serious issues that must soon be faced. One of these, perhaps the most important, is access to the broadest possible spectrum of genetic resources on which new cultivar development will depend. And, flowing from that, the conservation of these resources to ensure their continued and future availability. The cut flower industry requires a product which in general meets some of the following criteria; it should be a uniform, standardised bloom produced by clonal selection with a precisely predictable flowering time, marketed under a known recognised trade name and should be picked, packed, and graded to unvarying specifications. In other words, the buyer knows exactly what he is getting. However, such a cultivar represents an attenuated genetic base within a species or hybrid swarm. It serves the purpose for which it was raised, but it may not contain within its genetic complement a sufficient range of characters to be a good parent for a new range of cultivars having different characteristics.

To meet the requirement to generate new cultivars a plant breeder must have access to the largest possible pool of variation, preferably from wild sources. Without new and improved cultivars the protea industry will inevitably stagnate. About the only certainty in life is that nothing ever remains the same. Continuous, inexorable change is all we can confidently predict about the future. Add to this dimension those fickle mistresses, fashion and novelty and we see only too clearly why the continued breeding, selection, trial and release of new cultivars represent a prime driving force for the continued expansion of the industry. Markets are always looking for something new.

Changing fashion is especially important. Who can tell whether the cultivars being produced today will be desirable commodities on the European, American, or Japanese markets in thirty years time? Will an outsize pale pink *Protea cynaroides* still be a marketable product in the future or will some of the smaller darker forms of the species be more acceptable? We simply have no way of telling but in order to prepare for changing tastes and trends we must take steps to preserve the reservoirs of genetic variation within all the commercially important and potentially commercially important species. This of course applies not only to Southern Africa, but also to Australian Proteaceae and indeed any of the tropical African species, which might be used one day as breeding stock.

Variation manifests itself most obviously in the more widespread species but even those Cape Proteaceae with quite restricted distribution ranges show surprising degrees of variability between one local population and another. Form, stem length, shape and size of flower heads, colour and flowering time as well as less obvious qualities like disease resistance or adaptability to different substrates, are just a few of the characters that may differ quite significantly. Some examples of how rate or geographically restricted local races of widely dispersed species have been saved just in time and then used to develop new cultivars, will help to demonstrate my point.

> (a) *Leucospermum lineare* occurs in the Drakenstein mountains of the South Western Cape. The more widespread form has a rather sprawling habit, medium-sized inflorescences producing yellowish styles tipped with bright orange pollen presenters. At the extreme southern end of the species range in the French Hoek mountains near Robertsvlei, a more upright form occurs with larger inflorescences having uniformly bright orange styles. This rate and distinct local form has been almost entirely exterminated in nature as the natural habitat has been systematically planted to commercial pine forest during the past eighty years.

Fortunately, it was brought into cultivation in the 1940s by Kate Stanford, a pioneer protea grower who later supplied plants to Frank Batchelor. Duncan and Davies introduced this form into New Zealand in 1950, probably from Kate Stanford's seed (Mathews, 1983) Crossed with *Leucospermum cordifolium*, Gert Brits produced two outstanding hybrids in 1983 names Succession 1 and Succession 2.

(b) Leucospermum 'Starlight' is also a special selection of the red-flowered form of *Leucospermum lineare*. It was given to the Fynbos Research Unit by plant pathologist Kallie Petzer who found a few plants surviving on the edge of the Pine plantations in Assegaibosloof at French Hoek in the middle 1950s. This particular selection is characterised by a good strong scarlet flower colour and exceptionally long straight, purple-flushed stems. Pine plantations have eliminated the small population from which Petzer's long-stemmed, red-flowered selection was made, but the material, now fortuitously preserved in cultivation at Elsenburg, will ensure the perpetuation of its desirable characteristics.

The next example shows just how easy it is to lose important genetic resources when naturally occurring populations are destroyed.

(c) The story of Leucospermum 'Yellow Bird' and Leucospermum 'High Gold' began about twenty-five years ago in the late 1960s when Walter Middelmann accidentally discovered a small population of a prostrate growing form of *Leucospermum cordifolium* with clear yellow flowers growing on Botrivierplaas in the lower western foothills of Babylonstoring near Botrivier. Seed was collected and it was brought into cultivation on the Middelmann's farm where Gert Brits saw it flowering a few years later. He was given cuttings from this selection, which he named 'Yellow Bird'. 'Yellow Bird' was a considerable improvement on the original prostrate-growing wild type as it had a more upright habit. Even so, it still had a tendency to sprawl. A few years later in 1983 Gert Brits obtained flowering material of a yellow-flowered form of *Leucospermum patersonii* and crossed it with 'Yellow Bird'. Of the three seedlings that resulted from this cross the best has been named 'High Gold'. It has large rich yellow blooms, upright stems and tolerates slightly alkaline soils, a character it inherits from *Leucospermum patersonii*, which is endemic to limestone formations in the southern Cape. But most important is the previous yellow gene derived from the few individuals of prostrate growing L. cordifolium growing at Botrivier. In September 1993, an unsuccessful search for the original population of yellow coloured *Leucosperum cordifolium* on Botrivierplaas was undertaken by Mr. Middelmann's farm foreman who was well acquainted with its exact locality. The wild stock apparently no longer exists.

I hope that these examples will have convinced you that the conservation of genetic resources in the Southern African Proteaceae is pivotal to the development of improved cultivars and therefore to the long-term survival of the industry.

Before asking what can be done to promote the conservation of genetic resources in the Southern African Proteaceae, we have to consider the uncertain political and economic environment prevailing in the region at present.

South Africa is undergoing profound political and social change. Governments of the future will almost certainly modify the past patterns of state expenditure. For example, budgets for nature conservation may be greatly reduced in favour of housing and education. In consequence, the conservation of important gene pools in the wild could be seriously prejudiced. Moreover, the future of gene banks of cultivated material is equally problematic. Already the ARC, which maintains a proteaceous gene bank and funds the breeding program at Elsenburg has to rely less and less on state funding and will have to look for contracts or sponsorships to bolster future funding. These are some of the looming problems of the future. We cannot ignore them.

How then can the IPA use its influence and financial resources to secure what we might reasonably describe as a sound investment portfolio of genetic resources for future breeding?

Proposed action:

The conservation of genetic resources can be approached at a number of different levels and I would like to discuss some of these with you.

1. First and foremost is the need to ensure the continued conservation of as many wild populations in their natural habitats, as possible. In South Africa, the majority of these are in mountain catchments controlled by State conservation bodies while others are on private property. Why should we put so much effort

into conserving wild populations? It is precisely because most species of southern African proteaceae are presently represented in nature by a broad spectrum of variation in the form of local races, colour forms, or seasonal races, which flower at different times of the year that we still have an unparalleled opportunity to preserve this great reservoir of diversity and variation for the benefit of future breeding programmes. But this diversity may not be there forever. The uncontrolled spread of invasive pest plants, the changing political climate in South Africa in which state spending on nature conservation may be severely limited in favour of spending on social programmes could alter the picture. Also, the sheer force of exploitive human activities can be devastating. In this connection, some of you may recall the celebrated case in the mid 1980s where one farmer almost succeeded in wiping out the entire population of the highly decorative *Protea holosericeae* in the Worcester mountains. Here, with regard to habitat conservation I am suggesting that the IPA used its influence in persuading whatever government may be in power in the future to support the conservation of important natural resources like the proteaceae. If ethical and moral arguments don't work, perhaps economic arguments will. You are an international organisation, so never underestimate your influence.

2. The second level concerns identification and conservation of specific local populations. There is a priority in my view to catalogue and geographically pin-point as many wild populations as possible of all the commercially important and potentially commercially important species, documenting their horticultural properties and characteristics. Pioneering work in this respect was undertaken by Dr. Marie Vogts in the 1960s and 1970s when she enumerated the sometimes considerably variations between wild populations of some of the horticulturally important species like *Protea Cynaroides, P magnifica, P repens,* and *P Neriifolia.* Even today, few realise how far-sighted her work was, but as the data was contained in internal departmental reports its circulation was very restricted. Moreover, this initial survey was limited to proteaceae populations in State Forest Reserves. Consequently, a good deal of variation in populations on private property escapes detection. This work urgently need to be revised,

refined and extended especially in the commercially important species of the general *Protea, Leucadendron, Leucospermum,* and *Serruria*. There is a mistaken belief that we know all there is to know about the local races of Cape Proteaceae in the field. However, new discoveries continue to be made. May I remind you of the surprise discovery of a new *Mimetes, M crysanthus*, a few years ago. This species was discovered by a game guard in 1987 in the 9,500 hectare Gamka Mountain Nature Reserve near Oudtshoorn in the southern Cape. The reserve has been established in 1974 but thirteen years elapsed before the Cape Dept. of Nature Conservation discovered that their reserve harboured this magnificent new species. We cannot rule out further surprises or indeed the discovery of previously unknown variants of well-known species.

Earlier, I mentioned Leucadendron 'Safari Sunset'. Excellent as this is, it was produced using the best parents that were currently available in New Zealand. Careful screening of all the red-leaved forms of *Leucadendron salignum* from both the Langkloof and Koue Bokkeveld, would, I know, reveal superior potential parents for a similar cross (*Leucadendron laureolum x L. salignum*), which would almost certainly give rise to offspring with even clearer and richer colour tones than 'Safari Sunset'. However, to the best of my knowledge no one has thoroughly evaluated the wild populations of *L. salignum* to select the best reds for further breeding purposes. Fortunately, most of the red-leaved populations of *L. salignum* are not under serious threat at present and such a programme could quite easily be undertaken. It is this sort of material that should be documented and collected for further research.

I have spent much of my working life undertaking taxonomic revisions of the Southern African Proteaceae. *Surruria*, the last major genus to be revised is nearing completion. Vital as they are, taxonomic revisions simply don't provide data on variation at the level of individual populations, which is so essential to breeders and growers. As a start we need to document the horticultural characteristics of each wild population of the commercially important species as well as the soil types and microclimates of their habitat. Probably less than 50 species are involved so that task though considerable, is not insurmountable. This would be not only for immediate use, but to ensure the conservation of specifically earmarked populations as gene pools for further breeding programmes so as to meet the changing demands of the market in terms of style and fashion, as well as those less obvious characteristics such as disease resistance. Once identified these select local races could be brought into cultivation in existing gene banks. I believe a project such as this is well worth the support of organisation like the IPA.

1. The next level at which conservation action needs to be taken concerns the maintenance and expansion of existing gene banks, as for example the material cultivated by the South African ARC's Fynbos Research Unit at Elsenburg. Consider for instance the current cost of merely keeping up their unique collection of just of 1000 Proteaceae selections. Then plants of each of these 1000 selections are in cultivation — some 10,000 plants of breeding stock. Each plant costs R14 per year (US$5.0-0) which means it costs the South African tax payer R140,000 per year just to maintain this important gene bank even before any research is done with the material. This does not include major costs like salaries, equipment, and running costs. Projects such as Protea breeding will have to be increasingly self-supporting in the future as direct government funding is reduced.

   Even it if wished to support the maintenance of gene banks, the IPA may not have the resources to provide more than token support. Instead, as new cultivars are released to growers and producers it may be necessary to levy a small royalty on the material sold which could be ploughed back into maintaining the gene bank. Here, the IPA might serve a useful function as an international audit body for royalty collections.

2. Finally, efforts should be made to conserve genetic resources on a global scale.

3. In my view there is a need to develop gene banks of proteaceae breeding stock in each country, which intends to become a significant protea producer. Firstly, it would be prudent to provide insurance for the existing genetic material by spreading it globally. And, secondly gene banks of breeding stock specifically adapted to local climatic and edaphic conditions should be developed in

Israel, America, Australia, and New Zealand. Each growing area has its own peculiar requirements. For example, frost tolerance may be an important characteristic in New Zealand, but quite unimportant in South Africa. These different growing areas in different parts of the world will wish to conserve different variations within a species or hybrids; variations which suit their local conditions. Here again the IPA should use its influence to persuade agricultural authorities in these countries to establish gene banks of proteaceous material.

Conclusion

I hope I have been able to demonstrate to you today the importance of identifying, collecting, and conserving a wide range of genetic material within the Southern African Proteaceae so that a sustainable flow of new, better and more marketable products is assured for the future. We have already seen how genetic material, used in the development of fine new cultivars, was fortuitously rescued from wild populations, which have since disappeared due to various human factors. Every week, more wild populations are lost through environmental degradation. The process continues to accelerate.

I hope too that I have been able to highlight some of the threats and uncertainties that the future might hold for the conservation of these genetic resources both in nature and in gene banks. Moreover, I trust that the suggestions I have made regarding the IPA's possible financial and moral involvement will not cause offence. It is however clear that the industry must take steps to protect its own interests, and, while the opportunity still exists, make a serious commitment to conserve the genetic resources, or if you prefer, the raw material from which its future products will be built. I have no doubt that in a century from now there will be some stupendous new cultivars being sold on world markets, but their diversity and quality will depend on the determination and resolve of the present generation to protect the genetic capital which nature has bequeathed us.

APPENDIX K:

Safari Sunset

King Kiwi News, April 1981

Letters to the editor

The following letter was received in response to our report on the Leucadendron 'Safari Sunset' in the February issue of King Kiwi News — Editor.

Leucadendron Hybrids, Safari Sunset/Red Gem

Dear Sir,

In 1946 my in-laws, the late Wallace and Jean Stevens moved their rare plant nursery at Bulls to Wanganui, and concentrated on the winter cut flower and Iris side of their business. Their specialty flowers were the winter flowering African and Australian Proteaceae, i.e. Banksias, Proteas and the then unknown Leucadendrons, Leucospermums, Mimetes with others.

The red coloured form of *L. salignum* (formerly *L. adscendes*) was originally grown from seed obtained from a South African correspondent, Miss Marie Stegmanns, and was not particularly bright. In order to improve both colour and form, Mrs Stevens selected the best plants and began a long process of selected line breeding to produce a number of very bright female plants which we marketed in considerable quantities during the 1950s and 60s. Many other Leucadendrons were grown, including L. leureolum (L. decorum). Mrs Stevens was very occupied with her Iris breeding, and after having made the first of the known Leucadendron hybrids, *L. laureolum* (seed parent) x *L. salignum* (now known as *L. Red Gem*), deputized to me the task of improving and continuing hybridization. (I was then a partner with Mr & Mrs Stevens in the firm. My work continued for a number of years, and specimens of many were sent to Mr Ion Williams of South Africa, who has described them in his Revision of the Genus *Leucadendron* (published 1972). The hybrid now known as L. Safari Sunset was one of these hybrids, a selected seedling from a raising of 98 plants of L. salignum (seed parent) x L. laureolum. In addition to this female plant, a number of very good male plants were produced, and it is our intention to release one of these to Mr Lewis Matthews in the near future.

Leucadendron and Protea hybrids, including Red Gem and Safari Sunset, have been supplied by us to Wellington, Auckland, Christchurch, and Dunedin flower markets since the early 1960s. However rising freight costs, high city rates, coupled with declining prices on the N.Z. Markets and long spells of family illness, culminating in the deaths of both Mr & Mrs Stevens, have caused a long spell in hybrid raising.

Both Mrs Stevens and my hybrids were seen and photographed by South Africa's Mr W. Middelmann when he was in N.Z. early in the 1960s. Others to have seen these hybrids during the 1960s and early 1970s included the late Sir Victor Davies, Mr R. Barry of South Taranaki Nurseries, and Mr Gibson of Waitara.

I regret that some others of the very good early hybrids were not grown on by other florists or nurserymen. These have subsequently been lost by us and will now require careful selection and time to reproduce. Leucadendrons used by us in hybridizing have included L. *loranthifolium*, daphnoides, sessile, tinctum (round seeded); elimense, globosum (½ round seeded); laureolum, salignum, gandogeri, discolor, eucalyptifolium, macowanii, lanigerum, microcephalum, stelligerum (flat seeded).

Yours faithfully
Ian C. Bell,
Stevens Bros.,
Wholesale Florists.

www.ingramcontent.com/pod-product-compliance
Lightning Source LLC
Chambersburg PA
CBHW020735180526
45163CB00001B/244